Dissertation

Adaptive Wavelet and Frame Schemes for Elliptic and Parabolic Equations

Thorsten Raasch

2007

Bibliografische Information der Deutschen Nationalbibliothek

Die Deutsche Nationalbibliothek verzeichnet diese Publikation in der
Deutschen Nationalbibliografie; detaillierte bibliografische Daten sind
im Internet über http://dnb.d-nb.de abrufbar.

ISBN 978-3-8325-1582-9

Logos Verlag Berlin
Comeniushof, Gubener Str. 47,
10243 Berlin
Tel.: +49 030 42 85 10 90
Fax: +49 030 42 85 10 92
INTERNET: http://www.logos-verlag.de

Adaptive Wavelet and Frame Schemes for Elliptic and Parabolic Equations

Dissertation
zur
Erlangung des Doktorgrades
der Naturwissenschaften
(Dr. rer. nat.)

dem

Fachbereich Mathematik und Informatik
der Philipps-Universität Marburg

vorgelegt von

Thorsten Raasch
aus Frankenberg/Eder

Marburg/Lahn Februar 2007

Vom Fachbereich Mathematik und Informatik
der Philipps–Universität Marburg als Dissertation
angenommen am: 8. Februar 2007

Erstgutachter: Prof. Dr. Stephan Dahlke, Philipps–Universität Marburg

Zweitgutachter: Prof. Dr. Hans–Jürgen Reinhardt, Universität Siegen

Drittgutachter: Dr. Rob Stevenson, Utrecht University

Tag der mündlichen Prüfung: 14. März 2007

Acknowledgements

I would like to sincerely thank my research advisor, Professor Stephan Dahlke, whose guidance, patience and constant willingness for discussions about wavelet and frame analysis substantially helped me to finish this thesis.

I am also indebted to Professor Hans–Jürgen Reinhardt for being my second referee and in particular for guiding my first scientific steps related to the interesting field of inverse problems when I belonged to his research group.

At the same time, I am also deeply grateful to Rob Stevenson for his readiness to write the third referee report and for inspiring discussions on the numerical analysis of adaptive wavelet and frame methods.

Furthermore, I thank all the members of the AG Numerik/Wavelet–Analysis in Marburg for creating such a friendly atmosphere to work in. Special thanks go to Karsten Koch who has been much more than a colleague for me in the last years. I would like to thank also Manuel Werner for many fruitful discussions and for his assistance in some of the numerical experiments.

I also have to express my gratitude to the current and former members of Hans–Jürgen Reinhardt's research group who made my stay in Siegen a valuable time to remember.

Thanks also go to my former teacher, Heinrich Meier, for initially drawing my attention to the fascinating world of mathematics by his encouraging way of holding classes.

Finally, I would like to express my sincere gratitude to my parents whose support I could always rely on, regardless of the amount of time I spent at the computer when visiting them.

I also feel grateful to the Deutsche Forschungsgemeinschaft which financially supported the final phase of my stay in Marburg under Grant Da 360/7–1.

Zusammenfassung

In jüngster Zeit wurden sogenannte Waveletbasen erfolgreich in verschiedenen Gebieten der angewandten Mathematik und ihrer Nachbardisziplinen eingesetzt, so z.b. in der Signal– und Bildverarbeitung, aber auch bei gewissen Problemstellungen der Numerik wie etwa der Lösung von Operatorgleichungen.

Die Grundidee bei Wavelets besteht darin, Funktionensysteme zu betrachten, die sich im Wesentlichen durch Skalierung und Translation einer einzigen Funktion ergeben. Als besondere Vorteile gegenüber anderen Funktionensystemen haben sich die folgenden Schlüsseleigenschaften von Wavelets herausgestellt:

- Nach geeigneter Reskalierung bilden Wavelets eine Basis in verschiedenen wohlbekannten Funktionenräumen;

- Wavelets sind im Ort lokalisiert, im Gegensatz z.B. zum Fouriersystem;

- Wavelets besitzen verschwindende Momente, so dass innere Produkte mit glatten Funktionen verschwinden oder zumindest exponentiell mit wachsender Waveletskala abfallen.

Wir sind in der vorliegenden Arbeit insbesondere an numerischen Anwendungen interessiert. Hierin können alle der oben genannten Grundeigenschaften von Wavelets vorteilhaft ausgenutzt werden. So erlauben Wavelets durch ihre Approximationseigenschaften und ihre Lokalisierung im Ort die numerische Behandlung von elliptischen Randwertproblemen, wie z.B. der Poisson–Gleichung, im Rahmen eines Galerkin–Verfahrens. Im Gegensatz zu anderen Funktionensystemen sind Wavelets gleichfalls dazu geeignet, die Lösung einer Integralgleichung effektiv zu approximieren, da aufgrund der verschwindenden Momente die entsprechenden Galerkinsysteme gut durch dünn besetzte Matrizen angenähert werden können. Weiterhin gestatten Wavelets den Einsatz einfacher diagonaler Vorkonditionierer.

Um die numerische Simulation realistischer Probleme aus der Praxis überhaupt rechenbar zu machen, sind *adaptive* Methoden von besonderem Interesse. Diese passen die Diskretisierung mittels a posteriori Fehlerschätzern selbststeuernd an die unbekannte Lösung des Problems an. Seit 25 Jahren haben sich hierbei adaptive Finite–Element–Verfahren in der Praxis bewährt. Allerdings waren deren theoretische Konvergenzeigenschaften lange Zeit unklar, insbesondere die Frage der Optimalität. Im Gegensatz dazu haben verschiedene seit den 1990er Jahren entwickelte adaptive Wavelet–Methoden beweisbar optimale Konvergenz– und Komplexitätseigenschaften. Insbesondere im Fall symmetrischer elliptischer Probleme und deren Modifikationen ist der Einsatz von Wavelets mittlerweile gut verstanden.

Der aktuelle Stand der Forschung im Bereich adaptiver Wavelet–Methoden besitzt allerdings mehrere Schwachpunkte, von denen die folgenden beiden in der vorliegenden Arbeit behandelt werden sollen:

(P1) Üblicherweise lebt der betrachtete elliptische Operator auf einem beschränkten Gebiet oder einer geschlossenen Mannigfaltigkeit, so dass für die Umsetzung numerischer Wavelet–Verfahren eine entsprechende Wavelet–Konstruktion auf dem betreffenden Gebiet vonnöten ist. Alle bekannten Ansätze in diesem Bereich sind allerdings relativ kompliziert und liefern Wavelet–Systeme von zumeist unzureichender numerischer Stabilität. Hierdurch wurde bislang die Verwendung von Wavelet–Verfahren bei realistischen Problemen behindert.

(P2) Bislang ist weitgehend unklar, inwiefern sich adaptive Wavelet–Methoden auch zur numerischen Lösung nichstationärer Probleme wie z.b. parabolischen Anfangsrandwertproblemen eignen.

Zur Behebung des Problems (P1) diskutieren wir in dieser Arbeit den Einsatz sogenannter *Wavelet–Frames*. Hierbei handelt es sich um eine natürliche Verallgemeinerung des Begriffs der Rieszbasis, welcher üblicherweise einer Waveletbasis zu Grunde liegt. Um die Charakterisierung von Funktionenräumen auch im Fall von Frames sicher zu stellen, führen wir die Teilklasse der *Gelfand–Frames* ein. Diese erlauben es, in Analogie zu Wavelet–Rieszbasen, durch einfache Reskalierung des Gesamtsystems Frames in verschiedenen Funktionenräumen zu bilden. Um nun auf dem betrachteten beschränkten Gebiet geeignete (Gelfand–)Frames zu konstruieren, betrachten wir in Teil I eine überlappende Zerlegung des Gesamtgebiets in durch den Einheitswürfel parametrisierte Teilgebiete. Durch die Vereinigung geeignet gelifteter Referenzbasen auf dem Kubus erhält man auf einfache Weise einen globalen Wavelet–Frame. Die Grundeigenschaften der Refererenz–Waveletbasis wie Lokalität, Regularität und verschwindende Momente bleiben dabei erhalten. Allerdings ist der entstehende Frame redundant, d.h. die Entwicklungskoeffizienten einer gegebenen Funktion bezüglich des Frames sind nicht eindeutig. Zum Nachweis der Gelfand–Frame–Bedingung greifen wir auf die neuartige Theorie lokalisierter Frames zurück. In Teil II diskutieren wir den Einsatz von Gelfand–Frames bei der Diskretisierung elliptischer Operatorgleichungen. Analog zur Vorgehensweise bei Waveletbasen gestatten auch Frames eine äquivalente Darstellung der ursprünglichen Operatorgleichung in Framekoordinaten. Durch die Redundanz des Frames besitzt die biinfinite Systemmatrix hierbei einen nichttrivialen Kern, was den Einsatz von Galerkin–Methoden zunächst verhindert. Allerdings ist es stattdessen möglich, wohlbekannte lineare Iterationsverfahren auf den unendlich–dimensionalen Fall zu übertragen. Um ein implementierbares Verfahren zu erhalten, müssen dabei alle unendlich–dimensionalen Vektoren und Matrizen sowie deren Kombinationen durch hinreichend genaue endliche Approximationen ersetzt werden. Dieses ist in der Tat möglich unter Zuhilfenahme der Kompressionseigenschaften der verwendeten Wavelets und Wavelet–Frames. So kann zum Beispiel die adaptive Anwendung der biinfiniten Systemmatrix auf endliche Vektoren mit optimaler Komplexität durchgeführt werden. Ferner stehen für die Koeffizientendarstellung einer Iterierten implementierbare Thresholding–Routinen zur Verfügung. Durch die geeignete Kopplung dieser numerischen Grundbausteine geben wir eine auf Frames basierte inexakte Richardson–Iteration an und analysieren deren Konvergenz– und Komplexitätseigenschaften. Die theoretischen Ergebnisse werden durch ausgewählte numerische Testrechnungen illustriert. Für Problem (P2) geben wir schließlich in Teil III eine gangbare Strategie an. Wir befassen wir uns hier mit der Entwicklung adaptiver Wavelet–Methoden für lineare parabolische Gleichungen, wobei als Modellproblem die Wärmeleitungsgleichung mit einem Quellterm betrachtet wird. Inspiriert durch bereits etablierte Ansätze im Bereich Finiter Elemente geschieht die Diskretisierung des Gesamtproblems mit einem zweischrittigen Schema, der horizontalen Linienmethode. Zunächst wird mit Hilfe eines geeigneten Zeitintegrationsverfahrens das parabolische Anfangsrandwertproblem auf eine Folge elliptischer Probleme zurückgeführt. In dieser Arbeit betrachten wir hierzu linear–implizite Verfahren vom Rosenbrock–Typ. Die Ortsdiskretisierung der elliptischen Teilprobleme wird dann mit wohlbekannten adaptiven Wavelet–Methoden durchgeführt. Durch eine geeignete Kopplung der Zeitschrittweitensteuerung mit den Parametern des elliptischen Lösers erhalten wir ein voll adaptives Wavelet–Verfahren. Die Optimalität zumindest der Inkrementroutine kann gezeigt werden. Anhand numerischer Beispiele studieren wir zum Schluss die Eigenschaften des adaptiven Verfahrens.

x

Contents

Introduction

In recent years, wavelet bases have been successfully utilized for the solution of various problems in applied mathematics. Not only have they become a well–accepted tool in signal and image processing, wavelets have also been used in numerical analysis, especially for the treatment of elliptic operator equations.

From an abstract point of view, almost all wavelet bases share the fundamental idea to consider systems of functions generated by the dyadic dilates and integer translates of a single function like, e.g.,

$$\psi_{j,k}(x) = 2^{j/2}\psi(2^j x - k), \quad j, k \in \mathbb{Z},$$

which corresponds to a classical wavelet basis on the real line [66]. The particular advantages of wavelets are based upon their strong analytical properties. Among others, the following three features have emerged as the most important ones:

- Wavelet bases allow for the characterization of various smoothness classes, e.g., Sobolev, Hölder or Besov spaces, by weighted sequence norms of the corresponding wavelet coefficient arrays;

- Wavelets are localized in space in contrast to, e.g., the Fourier system;

- Wavelets have cancellation properties, meaning that the inner product between a smooth function and a wavelet either vanishes or decays exponentially as the scale of the wavelet increases.

In numerical applications, these key properties of wavelets can be exploited to a considerable extent. First of all, besides other systems of functions such as, e.g., finite elements, the density in classical smoothness spaces renders wavelets suitable for the discretization of elliptic boundary value problems like the Poisson equation on a bounded domain Ω in $\mathbb{R}^d$

$$-\Delta u = f \text{ in } \Omega, \quad u = 0 \text{ on } \partial\Omega.$$

By the spatial locality, the representation of differential operators with respect to a wavelet basis is at least quasi–sparse, i.e., the corresponding stiffness matrices in a wavelet–Galerkin discretization can be approximated well by finite sparse matrices. Fortunately, quasi–sparse representations in wavelet coordinates also hold for large classes of integral operators. This is in fact due to the cancellation properties of wavelets. Moreover, as a consequence of the equivalence between smoothness norms and weighted sequence norms of wavelet expansion coefficients, one obtains simple diagonal preconditioning strategies for the Galerkin system [55].

1

For the efficient numerical simulation of realistic problems from technical applications, *adaptive* approximation methods with a highly nonuniform spatial discretization are mandatory in order to keep the number of unknowns at a reasonable size. The core ingredient of most adaptive algorithms is an appropriate coupling of a posteriori error estimators and adaptive space refinement strategies in order to obtain reliable approximations of the unknown solution within prescribed error tolerances. For more than 25 years, adaptive finite element methods have been successfully used in practical applications. However, a full theoretical comprehension of their convergence properties, even for second–order elliptic problems, remained an open question for quite a long time [119]. In order to compare different adaptive schemes, it is a particular task to confirm that the method under consideration attains the best possible rate of convergence for the given input data. Concerning adaptive finite element discretizations of elliptic boundary value problems, optimal convergence rates for special refinement strategies and a relevant class of right–hand sides have been shown only recently [14, 135].

Since the late 1990s, particular interest has also been drawn to the analysis of adaptive discretization schemes based on wavelets. Contrary to finite element methods, the convergence properties of adaptive wavelet algorithms for a large range of problems have been apparent right from the start. By exploiting the analytic properties of wavelet bases, it was possible to design adaptive wavelet methods with guaranteed convergence for stationary symmetric elliptic problems, see [39, 46]. Shortly afterwards, it turned out that a specific variant of this approach is *asymptotically optimal* [33]. By this we mean that the number of unknowns needed to approximate the unknown solution up to a prescribed target accuracy asymptotically scales with the rate of the best N–term approximation as the target accuracy goes to zero. The applicability of the aforesaid class of wavelet methods to practical problems has been demonstrated by various numerical experiments in [7].

The motivations for this thesis arise from several deficiencies in the currently known theory and applications of adaptive wavelet methods. We shall explain these by looking at the status quo in this research area. By now, the application of wavelet methods to the adaptive numerical solution of operator equations is mainly guided by the following accepted principles. On the basis of the norm equivalences for wavelet bases, the original operator equation between Sobolev spaces can be rewritten as an equivalent discrete system over the wavelet coefficient sequence space. Add to this, a wavelet expansion of the current residual gives rise to reliable a posteriori error estimators. Coming from this initial point, there are two major strategies to derive adaptive wavelet schemes.

(I) Firstly, one may consider adaptive wavelet–Galerkin methods which implement an updating strategy that is steered by the large residual coefficients, leading to the algorithms considered in [33, 39, 46]. The computation of approximate residuals is feasible by exploiting the matrix compression properties of wavelet bases.

(II) A second approach, propagated in [34], is focused on the generalization of well–known iterative methods for finite–dimensional linear systems to the infinite–dimensional case. An approximate descent iteration of Richardson type was studied in [34, 82], where an asymptotically optimal method was obtained by carefully choosing the accuracies of the numerical subroutines. By their nature, descent iterations of the aforementioned type have generalizations towards nonsymmetric and nonstationary elliptic equations, but we will not go into further details here.

Both approaches share a critical bottleneck which we formulate as the first out of two major problems to be addressed in this thesis:

(P1) Usually, the elliptic operator under consideration is defined on a bounded domain or on a closed manifold, so that a construction of a suitable wavelet basis on this domain is needed. By now, there exist several constructions such as, e.g., [25, 26, 40, 62, 63, 95, 106, 136]. None of them, however, seems to be fully satisfactory in the sense that, besides the relevant positive virtues, these bases do not exhibit reasonable quantitative stability properties. Moreover, the constructions in the aforementioned references are all based on *non–overlapping* domain decomposition techniques, most of them requiring certain matching conditions on the parametric mappings. In practical situations, these may be difficult to satisfy. Finally, the handling of the overall wavelet basis in a computer program is quite subtle due to the complicated support geometry of the wavelets. By reason of this bottleneck, the potential of adaptive wavelet schemes has not been fully exploited in practice so far.

As a second motivation for this work, we would like to mention the following point:

(P2) So far, almost all known adaptive wavelet methods are designed to work for linear and nonlinear *stationary* problems. However, in many practical applications also time–dependent equations play an important role like, e.g., in heat conduction problems. It is then the question whether convergent adaptive wavelet methods can also be constructed for *nonstationary* problems and how the analytical properties of wavelets can be exploited in such a numerical scheme. Due to the close relationship to elliptic problems, a restriction of the problem class to linear boundary value problems of parabolic type seems advisable first.

This thesis will hence be focused on possible modifications and extensions of the aforementioned wavelet methods in order to provide possible solutions to both problems (P1) and (P2). In the sequel, we shall explain our targets in more detail.

Adaptive Frame Methods for Elliptic Equations

In order to solve problem (P1), one may consider an *overlapping* decomposition of the underlying domain $\Omega \subset \mathbb{R}^d$ into a union of smooth parametric images of the reference domain $\square = (0,1)^d$. The construction of a wavelet–like system on Ω then reduces to the lifting of a wavelet basis on the reference domain to the subdomains of Ω, followed by an aggregation of the local bases into a global system of functions [133]. Due to the overlap of the subdomains, one will not end up with a wavelet basis but with a redundant system of functions, a so–called (wavelet) *frame*. The concept of frames in a Hilbert space H has been introduced in [73], see also [29] for details. Compared with the case of wavelet bases, the construction of wavelet frames on bounded domains is drastically simplified. This is of particular importance when it comes to the realization of frame algorithms in computer software. Moreover, the approximation and cancellation properties of the reference wavelet basis on the cube immediately transfer to the global wavelet frame. For a systematical treatment of the arising principal questions when using frames instead of wavelet bases, let us formulate the following first task:

(T1) Given a polygonal domain $\Omega \subset \mathbb{R}^d$, provide a simple construction of wavelet frames $\Psi = \{\psi_\lambda\}_{\lambda \in \mathcal{J}}$ in $L_2(\Omega)$. Preferably, the frames should be able to characterize those function spaces required for the numerical discretization of linear elliptic operator equations.

On the other hand, due to the redundancy, the frame expansion coefficients of a given function are not unique, which has to be taken into account in numerical applications. To state an important consequence, adaptive frame–Galerkin approximations in the spirit of the aforementioned class (I) of adaptive wavelet methods prohibit themselves due to the fact that the corresponding Galerkin systems can be arbitrarily badly conditioned. Consequently, adaptive frame methods have to be either derived from the infinite–dimensional iterative schemes (II) or they need to be redesigned from scratch. It is most straightforward to consider the former class of methods for a start. As a second task to be addressed in this thesis, we therefore note the following program:

(T2) Generalize the ideas used for the wavelet discretization of elliptic operator equations to the case of frames. Analyze the mapping properties of the original elliptic operator in frame coordinates. Finally, formulate an approximate descent iteration that is guaranteed to converge in the case of a stationary symmetric elliptic problem. Similar to the algorithms using wavelet bases, the adaptive frame algorithm should be asymptotically optimal.

Adaptive Wavelet Methods for Parabolic Equations

As a second major topic of this thesis, we will address problem (P2), the application of adaptive wavelet methods to the numerical solution of linear parabolic boundary value problems. These can be written in the form of an abstract initial value problem in a Hilbert space H

$$u'(t) = Au(t) + f(t), \quad t \in [0, T], \quad u(0) = u_0,$$

where $A : D(A) \subset H \to H$ is a sectorial operator and $u : [0, T] \to H$ is the unknown solution. By the theory of analytic semigroups, the existence of a temporally smooth solution u is guaranteed, at least after an initial transient phase. In order to develop an adaptive wavelet scheme for the approximation of u, one may look at ideas developed in a finite element setting. In fact, based on the findings in [130, 131], adaptive wavelet methods for linear parabolic problems have been developed in the recent thesis [103]. The considerations in loc. cit. exploit the fact that the solution operators both for homogeneous and for inhomogeneous parabolic equations have a contour integral representation that allows for the construction of suitable adaptive quadrature rules. Though the results presented in [103] look promising, it is unclear at the moment whether the semigroup approach can be generalized to more complicated parabolic problems and we shall therefore choose another approach here.

Due to the initial value problem structure of the parabolic equation, it is a natural question whether some well–known techniques from the numerical discretization of *ordinary* initial value problems carry over also to the Banach or Hilbert space–valued case. In fact, for linear parabolic equations in a Hilbert space, implicit Runge–Kutta semidiscretizations in time have been studied and successfully applied in [17, 18, 19, 20], where the spatial discretization was done with finite elements. However, the Runge–Kutta approach is somewhat taylored to the treatment of *linear* problems. For a more general class of nonlinear problems, a Runge–Kutta semidiscretization in time will pose a number of quite expensive additional nonlinear equations per time step. So, despite the fact that we will not cover the discretization of truly nonlinear parabolic problems in this thesis, we argue in favor of a slightly different temporal discretization here. Precisely, we will focus on an S–stage *linearly* implicit semidiscretization in time, posing a system of s *linear* operator equations per time step. An error analysis for the unperturbed infinite–dimensional setting was firstly developed in [114], whereas the additional perturbation analysis in the case of a spatial approximation with finite elements can be found in [107, 108].

In this thesis, we are interested in the case where the spatial discretization uses wavelet bases. The analysis of wavelet methods for parabolic equations instead of, e.g., finite element methods, is done for several reasons which may justify in our opinion the higher computational work compared to finite elements. Firstly, due to the norm equivalences of a wavelet basis, an efficient preconditioning of the stage equations for a large range of time stepsizes will not be an issue for wavelet methods. Secondly, since we are going to solve the stage equations adaptively with well–known wavelet algorithms, we can rely on their optimal convergence and complexity behavior, drastically simplifying the analysis of the overall scheme. Moreover, using a fixed wavelet basis makes the linear combination of intermediate solutions from different stages painless. In a finite element discretization, one would have to interpolate between different grids ("mesh transfer") which might cause some technical difficulties. Summing up, we fix the work program concerning the application of wavelet methods to parabolic problems as the following task:

(T3) Develop a framework for adaptive wavelet discretizations of linear parabolic equations based on the aforementioned ideas. Analyze the convergence and computational complexity properties of the corresponding algorithms.

Layout

The thesis is structured as follows. In a preliminary Chapter 0, we shall briefly comment on the range of problems which the presented numerical schemes apply to. The remaining core of the thesis is then organized into three parts in accordance with the tasks (T1)-(T3) that have been listed above.

In Part I, we shall be concerned with task (T1), namely the construction of suitable wavelet frames on a polygonal domain $\Omega \subset \mathbb{R}^d$. The fundamental properties of wavelet bases are collected in Chapter 1. Among these are the Riesz stability and in particular the ability of characterizing various function spaces via weighted sequence norms of the wavelet coefficient arrays. We will also address the cancellation properties which will play a role in compressibility and localization arguments. In Chapter 2, we show then how to generalize the concept of wavelet Riesz bases with norm equivalences towards a frame setting. We shall end up with a subclass of frames, the so–called *Gelfand frames*. Inspired by the construction in [133], we will consider an *overlapping* domain decomposition of Ω into a union of smooth parametric images of the reference domain $\square = (0,1)^d$. By an appropriate lifting of a wavelet basis on the reference domain to the subdomains of Ω, and taking a union of these lifted bases, we obtain a global frame. Due to their nature, we refer to such frames as *aggregated wavelet frames*. In order to verify the Gelfand frame property for the constructed systems, we shall exploit the recently developed machinery of localized frames [77, 87, 89] to a considerable extent. The results presented in Chapter 2 have been published in [51, Ch. 6].

After the construction of suitable wavelet frames on the domain Ω, we will investigate the application of frames to the numerical discretization of elliptic operator equations in Part II, according to task (T2). In Chapter 3, the mapping properties of the elliptic operators under consideration are briefly reviewed. We recall results from both classical and non–classical regularity theory for the Poisson equation on Lipschitz and polygonal domains. For the theoretical analysis of adaptive wavelet methods, the smoothness of the target object in a specific scale of Besov spaces will be of particular interest. Chapter 4 is devoted to well–known results for linear and nonlinear wavelet approximation of a given function. In particular, it is recalled how the error of best approximation is related to the regularity properties of the target object. Moreover, we show how the compression properties of wavelet bases can be exploited to provide the building blocks of an adaptive wavelet scheme. Essentially, these consist of adaptive thresholding routines and the adaptive application of elliptic operators in wavelet coordinates. We also give two generic examples how implementable nonlinear wavelet approximation methods look like in practice. Based on the findings from the case of wavelet Riesz bases, we shall see in Chapter 5 that the fundamental subroutines of adaptive wavelet methods are also available in the case of frames. In order to obtain adaptive frame algorithms, we shall firstly be concerned with the mapping properties of the elliptic operator in frame coordinates. In fact, due to the redundancy of a frame, the biinfinite system matrix in a frame discretization has a nontrivial kernel. We shall show that this is not a problem for an exact infinite–dimensional Richardson iteration. It will turn out that an approximate version of the abstract scheme is indeed convergent, which can be achieved by judiciously choosing the tolerances of the numerical subroutines. However, since the iteration does produce kernel components of the current iterands, these have to be taken into account in the complexity analysis of the overall scheme. As a complement to the theoretical analysis, we shall give numerical examples in one and two spatial dimensions in order to validate the convergence and complexity properties of the adaptive frame schemes. The theoretical results of Chapter 5 can be found in [51].

Part III of this thesis is devoted to task (T3), the construction and numerical implementation of adaptive wavelet methods for the discretization of linear parabolic problems. In Chapter 6, we collect the basic properties of the linear parabolic problems under consideration. By using the theory of analytic semigroups, it is shown that the existence of a temporally smooth solution is guaranteed for a large class of problems. We address both the temporal and the spatial regularity properties of the unknown solution. By using results from the case of elliptic equations, it is possible to verify a high spatial Besov regularity also for the solutions of linear parabolic problems. Chapter 7 is devoted to the wavelet discretization of nonstationary problems. Based on the ideas mentioned above we shall employ a semidiscretization in time first. We briefly recall several details on the class of linearly–implicit integrators, in particular concerning convergence properties and the problem of local error estimation. In a second step, we perform a spatial discretization with wavelet Riesz bases. It will turn out that for a large range of stepsizes, the stage equations can be diagonally preconditioned quite efficiently due to the norm equivalences of the underlying wavelet basis. Moreover, by using the building blocks of adaptive wavelet schemes mentioned in Chapter 4, an adaptive increment algorithm can be specified that is asymptotically optimal. Finally, in Chapter 8, we present various numerical examples in one and two spatial dimensions in order to study the convergence and complexity behavior of the discussed wavelet schemes in practical situations.

Chapter 0

Range of Problems

In this preliminary chapter, we will comment on the class of operator equations which the presented analysis applies to. Despite the fact that many of the results addressed in the sequel also extend to the case of integral operators, we shall confine the setting to that of elliptic and parabolic boundary value problems on a bounded domain $\Omega \subset \mathbb{R}^d$ in either one or two spatial dimensions. The boundary of Ω is assumed to be piecewise linear, i.e., we are dealing with polygonal domains.

For both the elliptic and the parabolic equations we are interested in, we will consider a formally self–adjoint differential operator of order $2t = 2$ in divergence form

$$A(x, \partial) = \sum_{|\alpha|, |\beta| \leq 1} (-1)^{|\alpha|} \partial^\alpha (a_{\alpha,\beta}(x) \partial^\beta), \quad x \in \Omega, \tag{0.0.1}$$

where the coefficients $a_{\alpha,\beta}$ are assumed to be bounded and symmetric, $a_{\alpha,\beta} = a_{\beta,\alpha}$. Concerning elliptic equations, we are particularly interested in the numerical solution of boundary value problems with homogeneous Dirichlet boundary conditions and a right–hand side f

$$\begin{aligned} A(x, \partial)u(x) &= f(x) \quad \text{in } \Omega, \\ u(x) &= 0 \quad \text{on } \partial\Omega. \end{aligned} \tag{0.0.2}$$

The most prominent example is the *Poisson equation*

$$-\Delta u = f \text{ in } \Omega, \quad u|_{\partial\Omega} = 0, \tag{0.0.3}$$

which arises, e.g., in electrostatics, mechanical engineering and theoretical physics. The Poisson equation (0.0.3) will be the model problem in our numerical experiments of Chapter 5.

The corresponding bilinear form in a variational formulation of the boundary value problem (0.0.2) is given by

$$a(v, w) = \sum_{|\alpha|, |\beta| \leq 1} \int_\Omega a_{\alpha,\beta}(x) (\partial^\alpha v)(x) (\partial^\beta w)(x) \, \mathrm{d}x, \quad v, w \in H_0^1(\Omega). \tag{0.0.4}$$

Here $H_0^1(\Omega)$ denotes the usual Sobolev space with first order boundary conditions. For the Poisson equation (0.0.3), we obtain the bilinear form

$$a(v, w) = \int_\Omega \nabla v(x) \nabla w(x) \, \mathrm{d}x, \quad v, w \in H_0^1(\Omega). \tag{0.0.5}$$

In order to ensure that the weak formulation of (0.0.2) is well–posed, we will have to make further assumptions on a. For a closed subspace H of the Sobolev space $H^t(\Omega)$, we shall only consider those bilinear forms $a : H \times H \to \mathbb{C}$ that are *continuous*

$$\left| a(v,w) \right| \leq C_0 \|v\|_H \|w\|_H, \quad v, w \in H, \tag{0.0.6}$$

and H–*elliptic* in the sense that a is symmetric positive definite and

$$a(v,v) \gtrsim \|v\|_H^2, \quad v \in H. \tag{0.0.7}$$

Obviously, under assumption (0.0.6), the linear functional $a(v, \cdot)$ is continuous in H for any fixed $v \in H$. Therefore, given any Hilbert space $(V, \| \cdot \|_V)$ in which $H \hookrightarrow V$ is densely and continuously embedded, we can assume that $a(v, \cdot) \in V'$, where V' is the normed dual of V. As a consequence, a induces an operator $A : H \to H'$ by setting

$$\langle Av, w \rangle_{H' \times H} := a(v,w), \quad v, w \in H. \tag{0.0.8}$$

Here $\langle \cdot, \cdot \rangle_{H' \times H}$ refers to the duality pairing of H and H'. In the special case of the Poisson equation (0.0.3), the corresponding operator A shall also be referred to as the (negative) *Dirichlet Laplacian* $A = -\Delta_\Omega^D$ over Ω.

Combining (0.0.6) with the ellipticity condition (0.0.7), one readily infers that the variational problem

$$a(u,v) = f(v), \quad v \in H \tag{0.0.9}$$

has a unique solution $u \in H$ which depends continuously on $f \in H'$,

$$\|u\|_H \lesssim \|f\|_{H'}. \tag{0.0.10}$$

In other words, the operator equation

$$Au = f \tag{0.0.11}$$

is well–posed. This implication is also known as the Lax–Milgram theorem [92, 105], and it is equivalent to saying that the operator $A : H \to H'$ is boundedly invertible,

$$\|Au\|_{H'} \approx \|u\|_H, \quad u \in H. \tag{0.0.12}$$

Due to (0.0.7), the solution space H can also be equipped with the *energy norm*

$$\|v\|_a := a(v,v)^{1/2}, \quad v \in H. \tag{0.0.13}$$

Remark 0.1. *It should be noted that the restriction to symmetric bilinear forms a is done to guarantee convergence of those adaptive algorithms that are considered in the numerical experiments of Chapters 5 and 7. For the existence and uniqueness of a solution u to the variational problem (0.0.9), symmetry of the bilinear form a is not needed in general.*

Other examples covered by the assumptions on the bilinear form a are the Helmholtz equation $(A = -\gamma\Delta + I, \ 2t = 2)$ and the biharmonic equation $(A = \Delta^2, \ 2t = 4)$ over a bounded domain $\Omega \subset \mathbb{R}^d$. Moreover, this setting also covers operators of zero or negative order such as boundary integral equations over a closed manifold

$\Omega \subset \mathbb{R}^d$. Typical elliptic operators in this context are the single or double layer potential and the hypersingular operator. We refer to [55, 129] for further details. However, in this thesis we tacitly restrict the discussion to the case of elliptic differential operators of order $2t = 2$, although a more general operator order $2t$ is used in some of the referenced results.

The class of parabolic problems we are interested in is given by the temporally homogeneous initial–boundary value problem

$$\begin{aligned}
\tfrac{\partial}{\partial t}u(t,x) &= A(x,\partial)u(t,x) + f(t,x) && \text{in } (0,T] \times \Omega, \\
u(t,x) &= 0 && \text{on } (0,T] \times \partial\Omega, \\
u(0,x) &= u_0(x) && \text{in } \Omega.
\end{aligned} \tag{0.0.14}$$

Here $f : [0,T] \times \Omega \to \mathbb{R}$ is a suitable driving term. As the most prominent example, we mention the *heat equation*

$$\begin{aligned}
\tfrac{\partial}{\partial t}u(t,x) &= \Delta u(t,x) + f(t,x) && \text{in } (0,T] \times \Omega, \\
u(t,x) &= 0 && \text{on } (0,T] \times \partial\Omega, \\
u(0,x) &= u_0(x) && \text{in } \Omega.
\end{aligned} \tag{0.0.15}$$

The heat equation models diffusive processes in an isotropic and homogeneous medium, with applications, e.g., in physics, biology, mathematical finance and image processing. For the numerical experiments in Chapter 8, we will choose (0.0.15) as a model problem.

As we shall see in Chapter 6, the well–posedness of the parabolic problem (0.0.14) is again connected to the mapping properties of the differential operator $A(x,\partial)$ from (0.0.1), the induced bilinear form a and the corresponding operator $A : H \to H'$. By the continuous embedding $H \hookrightarrow H'$, the operator A will be viewed also as a linear unbounded operator from $(H, \|\cdot\|_{H'}) \subset H'$ onto H'. More generally, for any intermediate Hilbert space $(V, \|\cdot\|_V)$ such that $H \hookrightarrow V \hookrightarrow H'$, we will denote by

$$A_V := A|_{D(A;V)} : D(A;V) \subset V \to V \tag{0.0.16}$$

the *part of A in V*, where

$$D(A;V) := A^{-1}V = \{u \in H : Au \in V\} \tag{0.0.17}$$

is the *domain of A in V*. If the space V is fixed, we shall simply write A instead of A_V, which is justified since both operators coincide on $D(A;V)$. In particular, the interpretation of A acting on an intermediate space V will be used later in a Gelfand triple situation $H \hookrightarrow V \hookrightarrow H'$ with continuous and dense embeddings, where the intermediate pivot space V is identified with its normed dual V' via the Riesz mapping. Consequently, by the density of $V \simeq V'$ in H', we can assume that $D(A;V)$ is dense in H. Moreover, whenever $A : D(A;V) \subset V \to V$ is a closed operator, we can equip $D(A;V)$ with the *graph norm*

$$\|x\|_{D(A;V)} := \|x\|_V + \|Ax\|_V, \quad x \in D(A;V), \tag{0.0.18}$$

under which $D(A;V)$ is a Hilbert space. To simplify matters, we shall also abbreviate $D(A) := D(A;V)$ if V is the intermediate space of a Gelfand triple.

With these notational preparations and under the assumption that $A : D(A) \subset V \to V$ is a sectorial operator, see Chapter 6 for the concrete definitions, the parabolic problem (0.0.14) may then be considered as an abstract initial value problem

$$u'(t) = Au(t) + f(t), \quad t \in (0,T], \quad u(0) = u_0, \tag{0.0.19}$$

for a Hilbert space–valued variable $u : [0,T] \to V$. Let us remain a bit vague at the moment concerning the defining properties of what shall be considered a *solution* of (0.0.19). It will be convenient to write a solution u of (0.0.19) as a superposition $u = v + w$, where v solves the homogeneous problem

$$v'(t) = Av(t), \quad t \in (0,T], \quad v(0) = u_0 \tag{0.0.20}$$

and w solves the inhomogeneous problem

$$w'(t) = Aw(t) + f(t), \quad t \in (0,T], \quad w(0) = 0. \tag{0.0.21}$$

Part I

Wavelet Bases And Frames For
Operator Equations

Chapter 1

Wavelet Bases

This chapter is concerned with a brief overview about the basic properties and the construction of those wavelet bases that can be utilized in the numerical treatment of operator equations in a bounded domain Ω in $\mathbb{R}^d$. The results and strategies presented in this chapter are well–known in wavelet theory and can be found in basic textbooks [32, 143] or in the literature cited below.

To be precise, in Section 1.1, we will review the definition and some fundamental properties of wavelet–like Riesz bases. In Section 1.2, we show how it is possible to characterize various fundamental smoothness classes by the decay properties of the wavelet coefficients. Section 1.3 is devoted to the collection of some wavelet constructions on bounded domains that shall be used in the numerical examples later on.

1.1 Wavelet Riesz Bases

In the following, we consider a separable Hilbert space V with inner product $\langle \cdot, \cdot \rangle_V$ and induced norm $\|v\|_V := \langle v, v \rangle_V^{1/2}$. Moreover, we shall use a fixed countable, totally ordered index set $\mathcal{J}$. For any such index set and $p > 0$, let $\ell_p(\mathcal{J})$ be the space of all complex–valued sequences $\mathbf{v} = (v_\lambda)_{\lambda \in \mathcal{J}}$ over $\mathcal{J}$ such that $\|\mathbf{v}\|_{\ell_p}^p := \sum_{\lambda \in \mathcal{J}} |v_\lambda|^p$ is finite. Since $\mathcal{J}$ is fixed, we can abbreviate $\ell_p := \ell_p(\mathcal{J})$ in the following without possible confusion.

We shall firstly consider systems $\Psi = \{\psi_\lambda\}_{\lambda \in \mathcal{J}} \subset V$ that form a *Riesz basis* for V, i.e., any $f \in V$ has a unique expansion with coefficient array $\mathbf{c} = (c_\lambda)_{\lambda \in \mathcal{J}}$

$$f = \mathbf{c}^\top \Psi = \sum_{\lambda \in \mathcal{J}} c_\lambda \psi_\lambda \tag{1.1.1}$$

such that the following norm equivalence holds:

$$c_V \|\mathbf{c}\|_{\ell_2} \leq \|f\|_V \leq C_V \|\mathbf{c}\|_{\ell_2}. \tag{1.1.2}$$

Here $c_V, C_V \geq 0$ are called the *Riesz constants*.

Due to the estimate (1.1.2), the coefficient functionals $c_\lambda = c_\lambda(f)$ in the expansion (1.1.1) are bounded on V. Hence, by the Riesz representation theorem for linear bounded functionals on a Hilbert space, there exists a unique family of dual functions

$\tilde{\Psi} = \{\tilde{\psi}_\lambda\} \subset V$, such that $c_\lambda = c_\lambda(f) = \langle f, \tilde{\psi}_\lambda \rangle_V$. Moreover, it is $\|c_\lambda\|_{V'} = \|\psi_\lambda\|_V$, where V' is the normed dual of V and $\|g\|_{V'} := \sup_{\|f\|_V = 1} |g(f)|$ for all $g \in V'$. As a consequence of duality, the sets Ψ and $\tilde{\Psi}$ are *biorthogonal*, i.e., the relation

$$\langle \Psi, \tilde{\Psi} \rangle_V = \mathbf{I} \tag{1.1.3}$$

holds. This dual collection $\tilde{\Psi}$ is also a Riesz basis for V, with the Riesz constants C_V^{-1} and c_V^{-1}.

Note that in (1.1.3) and henceforth, we shall use the convenient shorthand notation

$$\langle \Theta, \Phi \rangle := \big(\langle \theta, \phi \rangle \big)_{\theta \in \Theta, \phi \in \Phi} \tag{1.1.4}$$

for any finite or infinite collection of functions $\Theta, \Phi \subset V$ and for any bilinear form $\langle \cdot, \cdot \rangle$ on $V \times V$. This abbreviation will also be used for one–element sets $\Theta = \{\theta\}$, so that the row vectors $\langle \theta, \Phi \rangle := \langle \{\theta\}, \Phi \rangle$ and the column vectors $\langle \Theta, \phi \rangle := \langle \Theta, \{\phi\} \rangle$ can be used without any confusion.

By the above argument, we see that biorthogonality of a system $\{\Psi, \tilde{\Psi}\}$ is necessary for the Riesz basis property to hold. But there is no equivalence, as a counterexample in [54, Section 3] shows. Consequently, additional structural properties of $\{\Psi, \tilde{\Psi}\}$ are needed that imply the Riesz stability. Here we are especially interested in the subclass of *wavelet* Riesz bases where the global index set $\mathcal{J}$ is decomposed as

$$\mathcal{J} = \bigcup_{j \geq j_0} \mathcal{J}_j, \quad \mathcal{J}_j := \big\{ \lambda \in \mathcal{J} : |\lambda| = j \big\}, \quad j \geq j_0, \tag{1.1.5}$$

with $j = |\lambda| \in \mathbb{Z}$ being the *level* or *scale* of a given index $\lambda \in \mathcal{J}$ and j_0 being some coarsest level. Alternatively, we can also write $\mathcal{J}$ as the limit of the index sets of all wavelet indices up to the level j

$$\mathcal{J}^j := \big\{ \lambda \in \mathcal{J} : |\lambda| \leq j \big\}, \quad j \geq j_0. \tag{1.1.6}$$

Besides the level j, a wavelet index λ will encode several further pieces of information on a single wavelet ψ_λ, e.g., the spatial location. In practice, λ will therefore almost always be some multiindex. We refer the reader to Section 1.3 for concrete examples.

For the following arguments, given any finite or infinite index set $\Lambda \subset \mathcal{J}$, we will denote by

$$\Psi_\Lambda := \{\psi_\lambda : \lambda \in \Lambda\} \tag{1.1.7}$$

the set of wavelets indexed by Λ and

$$S_\Lambda = \operatorname{clos}_V \Psi_\Lambda \tag{1.1.8}$$

shall be the V–closed span of them. Then, given a biorthogonal wavelet system $\{\Psi, \tilde{\Psi}\}$, the most commonly used general strategy for the verification of the Riesz basis property (1.1.2) utilizes further properties of the nested sequence of closed spaces

$$\mathcal{S} := (S_j)_{j \geq j_0}, \quad S_j := S_{\mathcal{J}^j}. \tag{1.1.9}$$

We will have to require specific approximation as well as regularity properties of the subspaces S_j, as becomes visible in the following fundamental theorem on basis–free Riesz stability:

Theorem 1.1 ([54, Th. 3.2]). *Assume that $\mathcal{Q} = (Q_j)_{j \geq j_0}$ is a sequence of uniformly bounded projectors $Q_j : V \to S_j$, such that*

$$Q_l Q_j = Q_l, \quad l \leq j. \tag{1.1.10}$$

Let $\tilde{\mathcal{S}} = \{\tilde{S}_j\}$ be the ranges of the sequence of adjoints $\mathcal{Q}' = (Q'_j)_{j \geq j_0}$. Moreover, suppose that there exists a family of uniformly bounded subadditive functionals $\omega(\cdot, t) : V \to \mathbb{R}^0_+$, $t > 0$, such that $\lim_{t \to 0+} \omega(f, t) = 0$ for each $f \in V$ and that the pair of estimates

$$\inf_{v \in V_j} \|f - v\|_V \lesssim \omega(f, 2^{-j}), \quad f \in V, \tag{1.1.11}$$

and

$$\omega(v_j, t) \lesssim \left(\min\{1, t2^j\} \right)^\gamma \|v_j\|_V, \quad v_j \in V_j, \tag{1.1.12}$$

holds for $\mathcal{V} = \mathcal{S}$ and $\mathcal{V} = \tilde{\mathcal{S}}$ with some $\gamma, \tilde{\gamma} > 0$, respectively. Then we have the norm equivalence

$$\| \cdot \|_V \approx N_{\mathcal{Q}}(\cdot) \approx N_{\mathcal{Q}'}(\cdot), \quad v \in V, \tag{1.1.13}$$

where

$$N_{\mathcal{Q}}(v) := \left(\sum_{j \geq j_0} \|(Q_j - Q_{j-1})v\|_V^2 \right)^{1/2}, \quad v \in V \tag{1.1.14}$$

and

$$N_{\mathcal{Q}'}(v) := \left(\sum_{j \geq j_0} \|(Q'_j - Q'_{j-1})v\|_V^2 \right)^{1/2}, \quad v \in V, \tag{1.1.15}$$

with $Q_{j_0-1} := Q'_{j_0-1} := 0$.

Estimates of the type (1.1.11) are also called *direct* or *Jackson type* estimates with respect to the *modulus* ω. They measure the approximation power of the underlying nested sequence of spaces S_j as $j \to \infty$. Conversely, estimates like (1.1.12) describe smoothness properties of the spaces S_j and are called *inverse* or *Bernstein type* inequalities.

Given a concrete biorthogonal wavelet system $\{\Psi, \tilde{\Psi}\}$, the original Riesz condition (1.1.2) is connected with the basis–free estimate (1.1.13) by the specific choice

$$Q_j v := \langle v, \tilde{\Psi}_{\mathcal{J}^j} \rangle \Psi_{\mathcal{J}^j}, \quad j \geq j_0. \tag{1.1.16}$$

Note that this operator $Q_j : V \to S_j$ is indeed a projector by the biorthogonality relation (1.1.3), and the idempotence condition (1.1.10) in Theorem 1.1 is automatically fulfilled. The adjoint of Q_j is given as

$$Q'_j v := \langle v, \Psi_{\mathcal{J}^j} \rangle \tilde{\Psi}_{\mathcal{J}^j}, \quad j \geq j_0, \tag{1.1.17}$$

and the property (1.1.10) guarantees that also the ranges $S'_j = \text{Ran}(Q'_j)$ are nested [46, 54].

In a concrete situation, we are therefore left to verify the sufficient conditions for the Riesz basis property in Theorem 1.1, more precisely the uniform boundedness of the operators Q_j and the validity of specific Jackson and Bernstein inequalities. In Section 1.3, we shall discuss how this is done in the case of various classical wavelet constructions. Before that, in Section 1.2, we will explain why Theorem 1.1 bears an even higher potential towards the simultaneous characterization of further function spaces besides V.

1.2   Characterization of Function Spaces

For the proof of the Riesz stability of Ψ in V, Theorem 1.1 required only that the Jackson inequality (1.1.11) and the Bernstein inequality (1.1.12) hold for some potentially small values $\gamma, \tilde{\gamma} > 0$. However, for many wavelet constructions we may expect that these exponents are indeed of a significant size. If, moreover, the modulus ω is chosen to be equivalent to the seminorm of some interesting subspace $H \subset V$, then the techniques employed in Theorem 1.1 also allow for an equivalent description of H based on the decay property of wavelet coefficients.

In the following, we shall hence review the definition of some classical function spaces and show how they can be characterized by weighted sequence norms of wavelet expansions. We confine the discussion to function spaces over a bounded Lipschitz domain $\Omega \subset \mathbb{R}^d$.

For $p \in (0, \infty]$, $L_p(\Omega) := L_p(\Omega; dt)$ shall be the space of all Lebesgue–measurable functions $f : \Omega \to \mathbb{C}$, such that the (quasi–)norm

$$\|f\|_{L_p(\Omega)} := \begin{cases} \left(\int_\Omega |f(x)|^p \, dx \right)^{1/p} & , \, p < \infty \\ \operatorname{ess\,sup}_{x \in \Omega} |f(x)| & , \, p = \infty \end{cases} \tag{1.2.1}$$

is finite. Analogously, $L_p(\Omega; \mu)$ shall be the L_p space related to the measure μ. For $p \geq 1$, the L_p spaces are Banach spaces, whereas for $p < 1$, they are only quasi–Banach spaces, since the triangle inequality will hold only up to a constant. The most important special case is $p = 2$, where $L_2(\Omega)$ is a Hilbert space with the inner product

$$\langle v, w \rangle_{L_2(\Omega)} := \int_\Omega v(x)\overline{w(x)} \, dx \tag{1.2.2}$$

and $\|v\|_{L_2(\Omega)}^2 = \langle v, v \rangle_{L_2(\Omega)}$. For a fixed domain Ω, we shall also use the abbreviation $\langle \cdot, \cdot \rangle := \langle \cdot, \cdot \rangle_{L_2(\Omega)}$ in the sequel.

Given a positive integer $m \in \mathbb{N}$, the *Sobolev space* $W^m(L_p(\Omega))$ is defined as the space of all functions $f \in L_p(\Omega)$ with weak partial derivatives $\partial^\alpha f$ in $L_p(\Omega)$ for all multiindices $\alpha \in \mathbb{N}_0^d$ with $|\alpha| = \alpha_1 + \cdots + \alpha_d = m$ and

$$|f|_{W^m(L_p(\Omega))} := \left(\sum_{|\alpha|=m} \|\partial^\alpha f\|_{L_p(\Omega)}^p \right)^{1/p}. \tag{1.2.3}$$

Under the norm $\| \cdot \|_{W^m(L_p(\Omega))} = \| \cdot \|_{L_p(\Omega)} + | \cdot |_{W^m(L_p(\Omega))}$, $W^m(L_p(\Omega))$ is a Banach space. The most important special case is again $p = 2$, where we get the Hilbert space $H^m(\Omega) := W^m(L_2(\Omega))$.

For fractional smoothness exponents $s \in (0, m)$, the corresponding Sobolev spaces $H^s(\Omega)$ can be introduced via the real interpolation method, see [1, 10, 11, 139]. To this end, given a pair of continuously and densely embedded Banach spaces $Y \hookrightarrow X$, one introduces the corresponding *Peetre K–functional*

$$K(f, t) := K(f, t; X, Y) := \inf_{g \in Y} \|f - g\|_X + t\|g\|_Y, \tag{1.2.4}$$

being continuous, non–decreasing and concave with respect to the t variable. Then, for parameters $\theta \in (0, 1)$ and $1 \leq q \leq \infty$, an intermediate Banach space $[X, Y]_{\theta, q}$ is given by the set of all functions $f \in X$ such that

$$\|f\|_{[X,Y]_{\theta,q}} := \|t^{-\theta}K(f,t))^q\|_{L_q(0,\infty;dt/t)} \tag{1.2.5}$$

is finite. If X, Y are Hilbert spaces, then so ist $[X, Y]_{\theta,q}$. Hence the L_2–Sobolev space for a general smoothness parameter can be defined as

$$H^s(\Omega) := [L_2(\Omega), H^m(\Omega)]_{s/m,2}, \quad s \in (0, m), \tag{1.2.6}$$

which coincides with the definition of Sobolev spaces for integer smoothness when $s \in \mathbb{N}$. For $s < 0$, we define $H^s(\Omega) := (H_0^{-s}(\Omega))'$ by duality.

Moreover, we shall also be concerned with the *Besov spaces* $B_q^s(L_p(\Omega))$ which arise as interpolation spaces between $L_p(\Omega)$ and $W^m(L_p(\Omega))$ [1, 69, 140]. For a more concrete definition of the corresponding Besov norm, one may use the *r–th L_p modulus of smoothness*

$$\omega_r(f,t)_{L_p(\Omega)} := \sup_{\|h\| \leq t} \|\Delta_h^r f\|_{L_p(\Omega_{rh})}, \quad t > 0. \tag{1.2.7}$$

Here, Δ_h^r is the r–th forward difference operator

$$\Delta_h^0 f := f, \quad \Delta_h^1 f := f(\cdot + h) - f, \quad \Delta_h^{k+1} := \Delta_h^1 \Delta_h^k, \tag{1.2.8}$$

and the admissible sets Ω_h are given by

$$\Omega_h := \left\{ x \in \Omega : x + th \in \Omega, \ t \in [0,1] \right\}, \quad h \in \mathbb{R}^d. \tag{1.2.9}$$

Then, for parameters $s > 0$ and $p, q \in (0, \infty]$, one can introduce the Besov spaces

$$B_q^s\big(L_p(\Omega)\big) := \left\{ f \in L_p(\Omega) : |f|_{B_q^s(L_p(\Omega))} < \infty \right\}, \tag{1.2.10}$$

where for $r := \lfloor s \rfloor + 1$ the Besov semi–(quasi–)norm is defined as

$$|f|_{B_q^s(L_p(\Omega))} := \begin{cases} \left(\int_0^\infty \left(t^{-s} \omega_r(f,t)_{L_p(\Omega)} \right)^q dt/t \right)^{1/q} & , \ 0 < q < \infty \\ \sup_{t \geq 0} t^{-s} \omega_r(f,t)_{L_p(\Omega)} & , \ q = \infty \end{cases}, \tag{1.2.11}$$

and $\| \cdot \|_{B_q^s(L_p(\Omega))} := \| \cdot \|_{L_p(\Omega)} + | \cdot |_{B_q^s(L_p(\Omega))}$ is the corresponding (quasi–)norm. Using equivalence results between $\omega_r(f, 2^{-j})_{L_p(\Omega)}$ and $K(f, 2^{-rj}; L_p(\Omega), W^r(L_p(\Omega)))$, cf. [101], it can be shown that Besov spaces are indeed interpolation spaces:

$$B_q^s(L_p(\Omega)) = [L_p(\Omega), W^r(L_p(\Omega))]_{s/r,q}, \quad s \in (0, r). \tag{1.2.12}$$

By the monotonicity of the modulus ω_r, a Besov space can also be endowed with the equivalent seminorm

$$\left\| (2^{sj} \omega_r(f, 2^{-j})_{L_p(\Omega)})_{j \geq 0} \right\|_{\ell_q(\mathbb{N})} \approx |f|_{B_q^s(L_p(\Omega))}. \tag{1.2.13}$$

Hence, for the L_2–modulus of smoothness $\omega = \omega_r(\cdot, t)_{L_2(\Omega)}$ which fulfills all the properties of a modulus as needed in Theorem 1.1, one can prove that $H^s(\Omega) \approx B_2^s(L_2(\Omega))$ for $0 < s < r$, see [70]. Using this equivalence, one obtains the following result on the characterization of Sobolev spaces which generalizes Theorem 1.1 in the case $V = L_2(\Omega)$.

Theorem 1.2 ([55, Th. 5.8]). *In the situation of Theorem 1.1 for* $V = L_2(\Omega)$, *assume that the direct estimate* (1.1.11) *holds in the form*

$$\inf_{v \in V_j} \|f - v\|_{L_2(\Omega)} \lesssim 2^{-sj} \|f\|_{H^s(\Omega)}, \quad f \in H^s(\Omega), \ 0 \le s \le m_V, \tag{1.2.14}$$

and that the inverse estimate (1.1.12) *holds in the form*

$$\|v_j\|_{H^s(\Omega)} \lesssim 2^{sj} \|v_j\|_{L_2(\Omega)}, \quad v_j \in V_j, \ s < \gamma_V, \tag{1.2.15}$$

for both scales of spaces $V \in \{S, \tilde{S}\}$, *where* $0 < \gamma := \min\{\gamma_S, m_S\}$ *and* $0 < \tilde{\gamma} := \min\{\gamma_{\tilde{S}}, m_{\tilde{S}}\}$, *respectively. Then we have the norm equivalence*

$$\|f\|_{H^s(\Omega)} \approx \Big(\sum_{j=0}^{\infty} 2^{2sj} \|(Q_j - Q_{j-1})f\|_{L_2(\Omega)}^2 \Big)^{1/2}, \quad s \in (-\tilde{\gamma}, \gamma). \tag{1.2.16}$$

Using again the special projectors Q_j from (1.1.16), we can restate (1.2.16) in terms of coefficient sequence norms as

$$\|f\|_{H^s(\Omega)} \approx \Big(\sum_{j=0}^{\infty} 2^{2sj} \sum_{|\lambda|=j} |\langle f, \tilde{\psi}_\lambda \rangle|^2 \Big)^{1/2}, \quad s \in (-\tilde{\gamma}, \gamma). \tag{1.2.17}$$

For the application of wavelet methods to the numerical solution of boundary value problems, it may be necessary that the corresponding boundary conditions are incorporated in the primal approximating spaces S_j. As a consequence, we can expect the direct estimate (1.2.14) to hold only on a proper subspace $H^s \subset H^s(\Omega)$. Then, at least for $s > -\frac{1}{2}$, the characterization results (1.2.16) and (1.2.17) can still be verified. For $s \le -\frac{1}{2}$, the situation is more complicated, in particular concerning wavelet bases on nontrivial bounded domains, see [55] for an extensive discussion.

It should be noted, however, that there are at least two situations where one definitely needs a characterization of the H^s subspaces with negative orders $s \le -\frac{1}{2}$. Firstly, those spaces may arise in the discretization of boundary integral operators like the single layer potential operator, and we refer to [129] for further details. Moreover, given some $f \in H^s(\Omega)$ with $s \ge 0$, we may need that the sequence of projections $(Q'_j f)_{j \ge j_0}$ onto $\tilde{S}_j$ converges to f in L_2 with a rate higher than $\frac{1}{2}$.

For those cases where one is also interested in a characterization of the H^s spaces with $s \le -\frac{1}{2}$, the dual basis has to be chosen appropriately. More precisely, the spaces $\tilde{S}_j$ have to fulfill *complementary* boundary conditions, meaning that $\tilde{S}_j$ (1.2.14) has to hold on a range of full Sobolev spaces $H^{\tilde{s}}(\Omega)$. Specific well–known wavelet constructions in this category from [61, 63, 126] are discussed in Section 1.3.

For a single wavelet $f = \psi_\lambda$, the norm equivalence (1.2.17) implies that we can control higher order smoothness norms of ψ_λ, since

$$\|\psi_\lambda\|_{H^s(\Omega)} \approx 2^{sj} \|\psi_\lambda\|_{L_2(\Omega)} \approx 2^{sj}, \quad \lambda \in \mathcal{J}. \tag{1.2.18}$$

Hence, given any subspace $H \hookrightarrow L_2(\Omega)$ with equivalent norm $\|\cdot\|_H \approx \|\cdot\|_{H^s(\Omega)}$ for some $s > 0$, we can infer from (1.2.17) that the inequality

$$c_H \|\mathbf{D}^s \mathbf{v}\|_{\ell_2} \le \|v\|_H \le C_H \|\mathbf{D}^s \mathbf{v}\|_{\ell_2}, \quad v = \mathbf{v}^\top \Psi \in H, \tag{1.2.19}$$

holds, where $c_H, C_H \geq 0$ are some constants and $\mathbf{D}$ is the diagonal matrix

$$(\mathbf{D})_{\lambda,\lambda} := 2^{|\lambda|}, \quad \lambda \in \mathcal{J}. \tag{1.2.20}$$

In other words, the *diagonally rescaled* basis $\mathbf{D}^{-s}\Psi$ is a Riesz basis in H. Up to a constant, in $(1.2.19)$ one may also replace the matrix $\mathbf{D}^s$ by the diagonal matrix $\mathbf{D}_H$ which consists of the H–norms of the single wavelets:

$$(\mathbf{D}_H)_{\lambda,\lambda} := \|\psi_\lambda\|_H, \quad \lambda \in \mathcal{J}. \tag{1.2.21}$$

By duality, one infers that every $v \in H'$ has an expansion $v = \tilde{\mathbf{v}}^\top \widetilde{\Psi}$ with $\tilde{\mathbf{v}} = \langle v, \Psi \rangle^\top$, so that

$$\|v\|_{H'} \approx \|\mathbf{D}^{-s}\tilde{\mathbf{v}}\|_{\ell_2}. \tag{1.2.22}$$

It is one of the crucial merits of wavelet bases that just by a diagonal rescaling, a wide range of Sobolev spaces including the L_2 case can be characterized. This is not possible, e.g., for other kinds of Riesz bases.

It should be noted that the characterization results from Theorems 1.1 and 1.2 have straightforward generalizations towards the case of reflexive Banach spaces like $L_p(\Omega)$ for $1 < p < \infty$. Using the characterization $(1.2.13)$ and a given L_2–biorthogonal wavelet basis with polynomial approximation order $m > s$ and of sufficiently high regularity, we can again employ the special projectors Q_j from $(1.1.16)$ to derive the characterization

$$\|f\|_{B_q^s(L_p(\Omega))} \approx \left(\sum_{j \geq j_0} 2^{jq(s+d(1/2-1/p))} \left(\sum_{|\lambda|=j} |\langle f, \widetilde{\psi}_\lambda \rangle|^p \right)^{q/p} \right)^{1/q}. \tag{1.2.23}$$

This equivalence can be shown to hold also for the case $p, q < 1$ given that $B_q^s(L_p(\Omega))$ is embedded in $L_1(\Omega)$, see [32], which is of particular importance in connection with nonlinear approximation. In particular, we shall be interested in characterizing the approximation spaces for best N–term approximation in H^t. As we will see in Section 4.1, these are exactly the Besov spaces $B_\tau^{sd+t}(L_\tau(\Omega))$, where τ and s are related via $\tau = (s + 1/2)^{-1}$. Inserting this special case into $(1.2.23)$ yields the important equivalence

$$\|f\|_{B_\tau^{sd+t}(L_\tau(\Omega))} \approx \left(\sum_{j \geq j_0} 2^{j\tau t} \sum_{|\lambda|=j} |\langle f, \widetilde{\psi}_\lambda \rangle|^\tau \right)^{1/\tau} = \|\mathbf{D}^t \langle f, \widetilde{\Psi} \rangle^\top\|_{\ell_\tau}, \tag{1.2.24}$$

where $\mathbf{D}^t \langle f, \widetilde{\Psi} \rangle^\top$ are exactly the expansion coefficients of f with respect to a Riesz basis $\mathbf{D}^{-t}\Psi$ in H^t.

1.3 Wavelet Bases on Bounded Domains

1.3.1 General Construction Principles

As already indicated in Section 1.1, wavelet bases in a Hilbert space V are typically constructed with the aid of a *multiresolution analysis*. By this we mean a system

$\mathcal{S} = (S_j)_{j \geq j_0}$ of linear, closed and nested subspaces that is asymptotically dense in V,

$$\mathrm{clos}_V \bigcup_{j=j_0}^{\infty} S_j = V. \tag{1.3.1}$$

The concept of a multiresolution analysis was originally introduced in a *shift–invariant* setting [116, 118]. There, the spaces S_j consist of functions on $\Omega = \mathbb{R}^d$, S_j is closed under integer translation $f \mapsto f(\cdot - k)$, $k \in \mathbb{Z}^d$, and the consecutive spaces S_j and S_{j+1} are related by *dilation*, i.e., $f \in S_j$ if and only if $f(2\cdot) \in S_{j+1}$. Due to $\Omega = \mathbb{R}^d$, the powerful tool of the Fourier transform may then be exploited for the construction of wavelet bases, see [37, 65, 66, 117, 118] for classical examples.

However, the setting of *bounded* domains Ω we are interested in clearly inhibits shift–invariance and the use of Fourier transform techniques. Nevertheless, some basic principles from classical wavelet theory still carry over to wavelet constructions for more general domain geometries. In the following, we will outline the basic steps for the design of a biorthogonal wavelet basis over a bounded domain Ω, as propagated, e.g., in [32, 59, 61]. Moreover, we assume from now on that $V = L_2(\Omega)$, which is sufficient for all cases of practical interest.

In a first step, one identifies stable *generator bases* $\Phi_j = \{\phi_{j,k} : k \in \Delta_j\}$ of the spaces S_j, so that

$$S_j = S(\Phi_j) = \mathrm{clos}_{L_2(\Omega)} \Phi_j, \quad j \geq j_0 \tag{1.3.2}$$

and likewise $\tilde{S}_j = S(\tilde{\Phi}_j)$, with a system of dual generators $\tilde{\Phi}_j = \{\tilde{\phi}_{j,k} : k \in \Delta_j\}$. In view of the later steps of the construction, we will have to require that the stability of the primal generators Φ_j holds uniformly in j

$$\|\mathbf{c}\|_{\ell_2(\Delta_j)} \approx \|\mathbf{c}^\top \Phi_j\|_{L_2(\Omega)}, \quad \mathbf{c} \in \ell_2(\Delta_j), \quad j \geq j_0. \tag{1.3.3}$$

In a shift–invariant setting, $\Delta_j = \mathbb{Z}^d$ is an infinite set and one would typically assume that the space S_0 is spanned by the integer translates $\phi_{0,k} = \phi(\cdot - k)$ of a single function ϕ,

$$S_0 = \mathrm{clos}_{L_2(\Omega)} \mathrm{span}\{\phi(\cdot - k) : k \in \mathbb{Z}^d\}. \tag{1.3.4}$$

Then the additional requirement (1.3.3) is automatically fulfilled for all $j \in \mathbb{Z}$ whenever one has stability in V_0, since in the shift–invariant setting, S_j is spanned by the integer translates

$$\phi_{j,k}(x) = 2^{-jd/2} \phi(2^j x - k), \quad k \in \mathbb{Z}^d, \ x \in \mathbb{R}^d \tag{1.3.5}$$

and an application of the transformation rule for $y = 2^j x$ traces back the ℓ_2–stability of Φ_j in V_j to that of Φ_0 in V_0.

On a bounded domain Ω, however, the index sets Δ_j will be finite and of increasing cardinality, as j tends to infinity. Moreover, Φ_j will not consist of the integer translates of a single function alone, so that the uniform stability (1.3.3) of Φ_j indeed has to be proved separately, see [54, 59]. A sufficient criterion for (1.3.3) to hold is the uniform boundedness of $\|\phi_{j,k}\|_{L_2(\Omega)}$ and $\|\tilde{\phi}_{j,k}\|_{L_2(\Omega)}$ in combination with *uniform locality* of the primal and dual generators, i.e., for a fixed constant $C < \infty$ we have

$$\#\{\mathrm{supp}\,\square_{j,k} \cap \mathrm{supp}\,\square_{j,k'} \neq \emptyset\} \leq C, \quad \mathrm{diam}\,\square_{j,k} \lesssim 2^{-j}, \tag{1.3.6}$$

where $\square_{j,k}$ is the smallest cube containing the supports of $\phi_{j,k}$ and $\tilde{\phi}_{j,k}$, respectively. The locality assumption is useful anyway in view of the numerical realization, and it can in turn be assured by some structural properties of the abstract nestedness condition, which shall be discussed next.

The inclusion $S_j \subset S_{j+1}$ connects the generators of two consecutive approximating spaces by the so–called *two–scale* or *refinement relation*

$$\Phi_j = \mathbf{M}_{j,0}^\top \Phi_{j+1}, \quad j \geq j_0. \tag{1.3.7}$$

Here $\mathbf{M}_{j,0} \in \mathbb{R}^{|\Delta_{j+1}| \times |\Delta_j|}$ is a matrix holding in its k–th column the expansion coefficients of $\phi_{j,k}$ with respect to the generators on the finer scale $j+1$. Since also the spaces $\tilde{S}_j$ are nested, the corresponding dual refinement matrices shall be denoted analogously by $\tilde{\mathbf{M}}_{j,0}$.

In a shift–invariant setting, the generator ϕ fulfills a special refinement equation of the form

$$\phi(x) = \sum_{k \in \mathbb{Z}} a_k \phi(2x - k), \quad x \in \mathbb{R}^d \tag{1.3.8}$$

with refinement coefficients $a_k \in \mathbb{R}$. For most of the practically relevant cases, one may assume that ϕ is compactly supported and that only finitely many a_k are nonzero. Hence $\mathbf{M}_{j,0}$ is a quasi–banded biinfinite matrix with entries $(\mathbf{M}_{j,0})_{k,l} = 2^{-1/2} a_{k-2l}$. Since the size and the entries of $\mathbf{M}_{j,0}$ are independent of the current refinement level in the shift–invariant case, $\mathbf{M}_{j,0}$ is then called *stationary*.

For wavelet constructions on a bounded domain Ω, the structure of the refinement matrices $\mathbf{M}_{j,0}$ will be more complicated. There, we can still hope that $\mathbf{M}_{j,0}$ is *quasi–stationary*, meaning that only the matrix dimensions change with j but, away from some level–independent corner blocks, $\mathbf{M}_{j,0}$ is still quasi–banded. This has the consequence that the interior generators will again be given as the dyadic dilates and translates of a single function, analogous to the shift–invariant case. From the quasi–bandedness of $\mathbf{M}_{j,0}$, we can hence easily infer the uniform stability (1.3.3) of the generators Φ_j. Moreover, quasi–stationary matrices $\mathbf{M}_{j,0}$ are *uniformly sparse*. By this we mean that the number of nonzero entries per row and column of these matrices remains uniformly bounded in j, which is useful in numerical computations. In Figure 1.1, the nonzero pattern of $\mathbf{M}_{j,0}$ is visualized, for the special case of a wavelet basis on the unit interval.

Moreover, the systems Φ_j and $\tilde{\Phi}_j$ shall be connected by the *biorthogonality* condition

$$\langle \Phi_j, \tilde{\Phi}_j \rangle = \mathbf{I}, \quad j \geq j_0. \tag{1.3.9}$$

This relation has several important consequences. By (1.3.3) and (1.3.9), also the stability of the dual system $\tilde{\Phi}_j$ will hold uniformly in j. Moreover, concerning the refinement matrices $\mathbf{M}_{j,0}$ and $\tilde{\mathbf{M}}_{j,0}$, the biorthogonality condition implies that $\mathbf{M}_{j,0}^\top \tilde{\mathbf{M}}_{j,0} = \mathbf{I}$ and we can introduce projectors Q_j and $\tilde{Q}_j$ onto the spaces S_j and $\tilde{S}_j$ via

$$Q_j f := \langle f, \tilde{\Phi}_j \rangle \Phi_j, \quad Q_j' f := \langle f, \Phi_j \rangle \tilde{\Phi}_j, \quad f \in L_2(\Omega). \tag{1.3.10}$$

Note that Q_j' is the L_2–adjoint of Q_j and these projectors will of course coincide with the projectors Q_j, Q_j' from (1.1.16) and (1.1.17). The uniform stability (1.3.3)

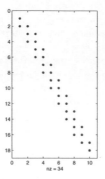

Figure 1.1: Nonzero pattern of the primal refinement matrix $\mathbf{M}_{j,0}$ for the spline wavelet basis on $\Omega = (0,1)$ from [126] with $m = 3$, $\tilde{m} = 3$, $j = 3$ and free boundary conditions.

of Φ_j corresponds to the uniform boundedness of the projectors Q_j, as needed in Theorem 1.1.

For the proof of the Riesz stability as well as for the characterization of function spaces, we need in particular specific approximation properties of the spaces S_j, $\tilde{S}_j$. To guarantee these, one will try to construct the generator bases Φ_j and $\tilde{\Phi}_j$ in such a way that they admit the reproduction of polynomials of order m and $\tilde{m}$, respectively,

$$\mathbb{P}_{m-1} \subset S_j, \quad \mathbb{P}_{\tilde{m}-1} \subset \tilde{S}_j, \quad j \geq j_0. \tag{1.3.11}$$

Here $\mathbb{P}_k$ denotes the set of all polynomials with total degree less or equal than k. In a shift–invariant setting, the inclusions (1.3.11) are meant in the sense of pointwise convergence. It can be shown that under the biorthogonality, stability and locality assumptions stated so far, the reproduction of polynomials (1.3.11) implies the validity of a Jackson estimate (1.2.14) for the values $m_S = m$, $m_{\tilde{S}} = \tilde{m}$, respectively, see [55]. In the case that Dirichlet boundary conditions of some order have to be incorporated into the primal multiresolution spaces S_j, the inclusion $\mathbb{P}_{m-1} \subset S_j$ is usually relaxed to merely hold in the interior of Ω, up to a boundary layer of thickness $c2^{-j}$.

Concerning the validity of the Bernstein estimate (1.2.15), we only note that due to the quasi–stationary refinement matrices, it is sufficient to verify that the generators Φ_j and $\tilde{\Phi}_j$ have a corresponding L_2–Sobolev regularity of order γ_S and $\gamma_{\tilde{S}}$, respectively, and we refer to [53] for the relevant proofs.

In the next step of constructing a wavelet basis over Ω, we have to pick some algebraic complement spaces W_j, $\tilde{W}_j$ such that

$$S_{j+1} = S_j \oplus W_j, \quad \tilde{S}_{j+1} = \tilde{S}_j \oplus \tilde{W}_j, \quad j \geq j_0. \tag{1.3.12}$$

Moreover, the complement spaces should be *biorthogonal* in the sense that

$$W_j \perp \tilde{S}_j, \quad \tilde{W}_j \perp S_j, \quad j \geq j_0, \tag{1.3.13}$$

which determines W_j and $\tilde{W}_j$ uniquely. Then, due to the nestedness of the spaces S_j and $\tilde{S}_j$, the operators $Q_j - Q_{j-1}$ and $Q'_j - Q'_{j-1}$ are projectors onto W_j and $\tilde{W}_j$, respectively.

Analogous to the generator sets Φ_j that span the spaces S_j, one then tries to find stable *wavelet* bases $\check{\Psi}_j = \{\check{\psi}_{j,k} : k \in \nabla_j\}$ for the spaces $W_j = S(\check{\Psi}_j)$. Again we assume here that the systems $\check{\Psi}_j$ are uniformly stable

$$\|\mathbf{c}\|_{\ell_2(\nabla_j)} \approx \|\mathbf{c}^\top \check{\Psi}_j\|, \quad \mathbf{c} \in \ell_2(\nabla_j), \quad j \geq j_0. \tag{1.3.14}$$

In view of the nestedness $W_j \subset S_{j+1}$, we can then express the wavelets on the level j with respect to the generator basis of the next higher scale

$$\check{\Psi}_j = \check{\mathbf{M}}_{j,1}^\top \Phi_{j+1}, \quad j \geq j_0, \tag{1.3.15}$$

where $\check{\mathbf{M}}_{j,1} \in \mathbb{R}^{|\Delta_{j+1}| \times |\nabla_j|}$. Moreover, it turns out that the construction of uniformly stable complement bases $\check{\Psi}_j$ is equivalent to *completing* the refinement matrices $\mathbf{M}_{j,0}$ to invertible mappings

$$\check{\mathbf{M}}_j = (\mathbf{M}_{j,0}, \check{\mathbf{M}}_{j,1}) : \ell_2(\Delta_j) \oplus \ell_2(\nabla_j) \to \ell_2(\Delta_{j+1}), \tag{1.3.16}$$

where the operator norms of $\check{\mathbf{M}}_j$ and of the inverse $\check{\mathbf{G}}_j := \check{\mathbf{M}}_j^{-1}$ stay uniformly bounded in j. This construction principle is known as the method of *stable completion* [28]. It is useful to block also $\check{\mathbf{G}}_j$ with

$$\check{\mathbf{G}}_j = \begin{pmatrix} \check{\mathbf{M}}_{j,0}^\top \\ \check{\mathbf{G}}_{j,1}^\top \end{pmatrix}, \tag{1.3.17}$$

where $\check{\mathbf{G}}_{j,1}$ contains in its columns the expansion coefficients of some dual wavelet basis $\check{\tilde{\Psi}}_j$ for $\tilde{W}_j$ with respect to $\tilde{\Phi}_{j+1}$. The invertibility of $\check{\mathbf{M}}_j$ then implies the matrix equation

$$\mathbf{I} = \check{\mathbf{M}}_j \check{\mathbf{G}}_j = \mathbf{M}_{j,0} \check{\mathbf{M}}_{j,0}^\top + \check{\mathbf{M}}_{j,1} \check{\mathbf{G}}_{j,1}^\top. \tag{1.3.18}$$

It should be noted that (1.3.18) is not uniquely solvable. In particular, it is not a priorily clear whether there exist *biorthogonal* wavelet bases, i.e., whether one may find special wavelet bases $\Psi_j = \mathbf{M}_{j,1}^\top \Phi_{j+1}$, $\tilde{\Psi}_j = \tilde{\mathbf{M}}_{j,1}^\top \tilde{\Phi}_{j+1}$ that satisfy

$$\langle \Psi_j, \tilde{\Psi}_j \rangle = \mathbf{I} \tag{1.3.19}$$

and that are moreover also uniformly locally supported. The delicate problem here is that one has to determine a sparse matrix $\mathbf{M}_j$ of the above form such that its inverse $\mathbf{G}_j$ is also sparse, which is nontrivial. It could be shown in [28] how the solution space of (1.3.18) may be parametrized by a family of linear transformations as soon as one initial stable completion $\check{\mathbf{M}}_{j,1}$ is known. In the latter case, equation (1.3.18) has exactly one biorthogonal solution

$$\mathbf{M}_{j,1} = (\mathbf{I} - \mathbf{M}_{j,0} \tilde{\mathbf{M}}_{j,0}^\top) \check{\mathbf{M}}_{j,1}, \tag{1.3.20}$$

and the corresponding dual block in $\mathbf{G}_j$ can be computed as

$$\tilde{\mathbf{M}}_{j,1}^\top := \mathbf{G}_{j,1}^\top = (\mathbf{I} + \tilde{\mathbf{M}}_{j,0}^\top \mathbf{M}_{j,1}) \check{\mathbf{G}}_{j,1}^\top. \tag{1.3.21}$$

Moreover, if the initial stable completion is uniformly locally supported, then this also holds for the biorthogonal completion (1.3.20). By (1.3.19), we can infer the matrix equation $\mathbf{M}_{j,1}^\top \tilde{\mathbf{M}}_{j,1} = \mathbf{I}$.

The overall wavelet Riesz basis Ψ is obtained by aggregating the single complement bases Ψ_j and the generators Φ_{j_0} from the coarsest level. In order to write this down in the most convenient way, we will use multiindices $\lambda = (j, e, k)$ with an additional *type parameter* $e \in \{0, 1\}$, and we set

$$\psi_{(j,0,k)} := \phi_{j,k}, \quad k \in \Delta_j, \quad \psi_{(j,1,k)} := \psi_{j,k}, \quad k \in \nabla_j. \tag{1.3.22}$$

The overall wavelet index set $\mathcal{J}$ is then given by

$$\mathcal{J} := \{j_0\} \times \{0\} \times \Delta_{j_0} \cup \bigcup_{j \geq j_0} \{j\} \times \{1\} \times \nabla_j, \tag{1.3.23}$$

so that we can define $\Psi := \{\psi_\lambda\}_{\lambda \in \mathcal{J}}$. Analogous abbreviations shall be used for the dual wavelets $\tilde{\psi}_\lambda$. For each $\lambda = (j, e, k) \in \mathcal{J}$, we denote by $|\lambda| = j$ the corresponding scale and by $e(\lambda) = e$ the type.

1.3.2   Cancellation Properties

In case that the dual generator bases $\tilde{\Phi}_j$ can reproduce polynomials of order $\tilde{m}$ as in (1.3.11), then we infer from the biorthogonality relation (1.3.13) that the primal wavelets have *vanishing moments* of order $\tilde{m}$,

$$\langle x^i, \psi_\lambda \rangle = 0, \quad 0 \leq i \leq \tilde{m} - 1, \quad \lambda \in \mathcal{J}, \ e(\lambda) \neq 0. \tag{1.3.24}$$

Analogously, the dual wavelets will have vanishing moments of order m if Φ_j reproduces polynomials of order m. As a consequence of (1.3.24), we can derive a so–called *cancellation property* of order $\tilde{m}$. By this we mean that inner products of ψ_λ with smooth functions decay exponentially fast as the scale $j = |\lambda|$ tends to infinity as

$$\left| \langle f, \psi_\lambda \rangle \right| \lesssim 2^{-|\lambda|(\frac{d}{2} + \tilde{m})} |f|_{W^{\tilde{m}}(L_\infty(\mathrm{supp}\,\psi_\lambda))}, \quad \lambda \in \mathcal{J}, \ e(\lambda) \neq 0. \tag{1.3.25}$$

In the more general case that the dual multiresolution spaces $\tilde{S}_j$ are only *accurate* of order $\tilde{m}$, meaning that the Jackson inequality (1.2.14) is valid for $m_{\tilde{S}} = \tilde{m}$ without using polynomial exactness arguments, we can still expect that (1.3.25) holds.

   The annihilation of polynomials by wavelets can be exploited in the localization theory of frames which is discussed in Chapter 2. Moreover, vanishing moment or cancellation properties play a crucial role in matrix compression and in the adaptive application of certain differential and integral operators in wavelet representation, as we shall see in Chapter 4.

1.3.3   Wavelet Constructions on the Interval

After having sketched the abstract strategy of wavelet constructions on bounded domains, we shall now specify concrete examples for the special case of the unit interval $\Omega = (0,1)$. In view of the application of these bases in the numerical treatment of operator equations, some obstructions arise which shall be listed first.

On the one hand, we are particularly interested in those biorthogonal wavelet bases that admit the incorporation of Dirichlet boundary conditions into the primal multiresolution spaces S_j, while the dual spaces $\tilde{S}_j$ should fulfill *complementary* boundary conditions, if possible. The bases should be able to characterize the appropriate scales of Sobolev and Besov spaces that are needed in the numerical examples. Moreover, at least the primal wavelets ψ_λ and their derivatives should be accessible analytically, making arbitrary point evaluations and quadrature cheap.

These requirements already exclude various classical wavelet constructions on the interval. Just to state some examples, *periodic* wavelet bases are useless for our purpose since they do not fulfill the appropriate boundary conditions nor do they reproduce the correct set of polynomials. In *orthogonal* wavelet constructions like [38], the primal wavelets are only given implicitly which makes them difficult to handle numerically. Moreover it is neither possible to increase the number of vanishing moments of the primal wavelets independent from the order of accuracy, nor can any orthogonal wavelet construction realize complementary boundary conditions. For the same reasons, the orthogonal spline *multiwavelet* bases from [71, 72] cannot be used. Though the latter constructions look interesting for their relatively high order of accuracy, their numerical realization seems not easy. Other biorthogonal multiwavelet bases on the interval were constructed in [56], where the primal generators are the cubic Hermite interpolatory splines. Here it is not fully clear at the moment whether also complementary boundary conditions can be realized with the primal and dual generators. Further constructions one may look at use so–called *prewavelets*. There, the complement spaces W_j are mutually orthogonal with respect to the L_2 or another energy inner product, drastically simplifying the stability analysis. The construction of prewavelets is relatively easy, see, e.g., [30, 100], but the dual wavelets are often not explicitly known and they are globally supported. In the case that the prewavelet spaces W_j are chosen orthogonal with respect to another inner product than that of L_2, the dual generators and wavelets may moreover have only a low Sobolev regularity and they mey even fail to belong to L_2, as is the case for the dual prewavelet basis from [100]. Especially from the viewpoint of tight Riesz bounds, also the construction of spline wavelets on the interval in [15] looks quite interesting, though the corresponding dual generators are unknown so far.

In principle, the interval bases we shall consider in this thesis are based on the constructions in [59, 61, 126]. There, the primal generators $\phi_{j,k}$ live in a spline space of order m with respect to some knot sequence $\{t_k^j\}_k \subset [0,1]$. In the interior of the domain $\Omega = (0,1)$, the primal generators in the mentioned constructions coincide with dilates and translates of cardinal B–splines of order m, up to a factor. As a consequence of the embedding into spline spaces, the critical L_2–Sobolev regularity of the primal multiresolution spaces is given by $m - \frac{1}{2}$.

For the verification of the polynomial exactness of order m, some modifications of the generators at the boundaries are necessary. Either one uses specific linear combinations of restricted cardinal B–splines [59, 61] or one resorts to B–splines related to the *Schoenberg knot sequence* [126] with m–fold boundary knots,

$$t_k^j = \min\left\{1, \max\{0, 2^{-j}k\}\right\}, \quad k = -m+1, \ldots, 2^j + m - 1, \tag{1.3.26}$$

where the corresponding B–splines of order m are defined as

$$B_{k,m}^j(x) := (t_{k+m}^j - t_k^j)[t_k^j, \ldots, t_{k+m}^j]_t(t - x)_+^{m-1}. \tag{1.3.27}$$

Here $(x)_+ := \max\{0, x\}$ and $[t_0, \ldots, t_k]_t$ denotes the divided difference operator corresponding to the knots t_i acting in the t variable. In the construction of [126], the primal spline generators of order m are then simply defined as $\phi_{j,k} := 2^{j/2} B_{k,m}^j$. Examples of these generators in the case $m = 3$ can be found in Figure 1.2.

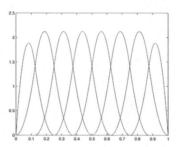

Figure 1.2: Primal spline generators $\phi_{j,k}$ from [126] with $m = 3$, $j = 3$ and homogeneous Dirichlet boundary conditions.

In any of the mentioned spline wavelet constructions, the primal refinement matrices $\mathbf{M}_{j,0}$ can be computed by simply solving some triangular systems of linear equations, see [59, 126] for details.

The recent approach from [126] seems favorable for several reasons. Firstly, the primal multiresolution spaces S_j coincide with the full spline spaces related to the knot sequences $\{t_k^j\}_{k=-m+1}^{2^j+m-1}$, which is not always the case for the construction in [59]. Moreover, the incorporation of homogeneous Dirichlet boundary conditions of some order into the primal multiresolution spaces is easily done by omitting the corresponding number of boundary knots. Point evaluation of the primal generators and their derivatives is painless since B–splines fulfill numerous recurrence relations that allow for fast and stable evaluation algorithms [67]. Add to this, using only B–splines as primal generators drastically simplifies the computation of supports and singular supports in [126] compared with the constructions in [59]. Finally, choosing the shortest possible supports of the boundary generators seems to have a favorable impact on the L_2–Riesz constants of the wavelet system.

In order to realize complementary boundary conditions for the multiresolution spaces S_j and $\tilde{S}_j$, the dual generators $\tilde{\phi}_{j,k}$ are chosen to be exact of the full order $\tilde{m}$ and to be biorthogonal to the primal generators. Since the concrete construction is quite subtle, we refer the reader to [59, 61, 126] for the technical details. Let us only note that the dual refinement matrices $\tilde{\mathbf{M}}_{j,0}$ can again be set up by solving several triangular systems, see [59, 126] for examples. Generally speaking, in all known biorthogonal spline wavelet constructions on the interval, the dual generators are given as linear combinations of some dual scaling functions from the shift–invariant construction [37]. As a consequence, also the critical L_2–Sobolev regularity of the dual generators is known and can be computed with well-known algorithms[1] [99], see Table 1.1 for various example values.

m	$\widetilde{m}$	$\nu_2(\widetilde{\phi})$
2	2	0.44076544507035
2	4	1.17513151026734
2	6	1.79313390050964
3	3	0.17513151026735
3	5	0.79313390050989
3	7	1.34408387241967
3	9	1.86201980785003
4	6	0.34408387241950
4	8	0.86201980784985
4	10	1.36282957312823

Table 1.1: Critical L_2–Sobolev regularity of some dual scaling functions from [37].

Some specific dual generators from [126] are given in Figure 1.3. However, it should be noted that at least in the numerical discretization of elliptic operator equations with wavelets, the dual generators and wavelets are never needed explicitly. There they merely serve to ensure the corresponding number $\widetilde{m}$ of vanishing moments for the primal wavelets ψ_λ and to establish an appropriate negative lower bound $-\widetilde{\gamma}$ in (1.2.16).

Concerning the construction of the primal wavelets, the approaches [59, 62, 63, 126] utilize the method of stable completion, as introduced in the previous subsection. In [59] it has been shown how an initial stable completion can immediately be derived from a specific factorization of $\mathbf{M}_{j,0}$. After performing the biorthogonalization step (1.3.20), the interior wavelets coincide with those presented in [37], whereas the boundary wavelets look more complicated. However, the factorization algorithm used in [59] does not automatically lead to the *maximal* number of interior wavelets. This is of particular importance since the Riesz constants of the overall wavelet basis crucially depend on the shape of the boundary functions. Moreover, due to the small supports of the interior wavelets, the sparsity of stiffness matrices is potentially higher. In [126], a refined algorithm was developed that produces the optimal number of interior wavelets. Figure 1.4 shows some primal wavelets from the latter construction.

From the viewpoint of numerical stability, in particular concerning tight L_2– and H^s–Riesz bounds, the wavelet bases from [126] clearly outperform those of the original construction in [59, 63]. It should be mentioned that in [9], the latter wavelet bases have been substantially stabilized by orthogonalizing the boundary wavelets with respect to some energy inner product. However, a comparison with the benchmark values in [58] reveals that the Riesz constants as well as the spectral

---

[1]A software package for the estimation of critical Sobolev exponents of (multi)wavelets can be found on the website http://www.mathematik.uni-marburg.de/~dahlke/ag-numerik/research/software.

properties of stiffness matrices from a wavelet–Galerkin discretization of elliptic differential equations are still favorable in [126], as long as the spline order is restricted to low values $m \leq 3$. As indicated by the recent construction [15], future research in the direction of high order biorthogonal spline wavelets on the interval remains potentially fruitful.

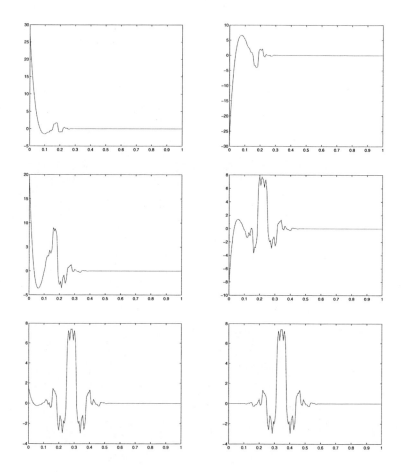

Figure 1.3: Some dual spline generators $\tilde{\phi}_{j,k}$ from [126] with $m = 3$, $\tilde{m} = 5$, $j = 4$ and complementary boundary conditions.

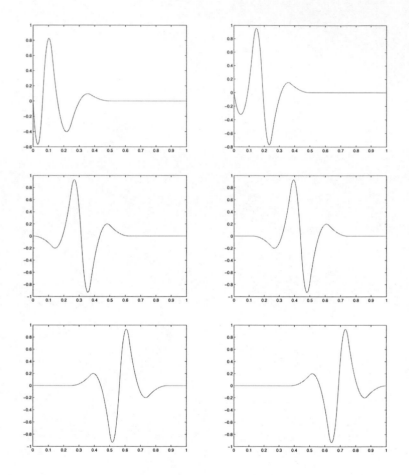

Figure 1.4: Some primal spline wavelets $\psi_{j,k}$ from [126] with $m = 3$, $\tilde{m} = 3$, $j = 3$ and complementary boundary conditions.

1.3.4 Wavelet Constructions on Polygonal Domains

On more general domains $\Omega \subset \mathbb{R}^d$ with $d \geq 2$, there are several possible wavelet constructions available, most of them being based on domain decomposition ideas. The domain Ω under consideration is then subdivided into a disjoint union of M parametric images of the unit cube $\Box = (0,1)^d$

$$\overline{\Omega} = \bigcup_{i=1}^{M} \overline{\Omega}_i, \quad \Omega_i := \kappa_i(\Box), \quad i = 1, \dots, M, \tag{1.3.28}$$

where the subpatches Ω_i only meet along lower–dimensional surfaces and the charts κ_i are assumed to be sufficiently smooth. Examples of wavelet constructions based on the decomposition (1.3.28) can be found in [24, 25, 26, 40, 62, 63, 95, 136]. In this thesis, we shall mainly focus on the construction of *composite wavelet bases* in [62] and we will hence recall some basic features thereof in the sequel.

Any of the constructions for quadriliteral decompositions of Ω utilizes an auxiliary wavelet basis $\Psi^{\Box} = \{\psi_\lambda^{\Box}\}_{\lambda \in \mathcal{J}^{\Box}}$ on the unit cube $\Box$ with dual basis $\tilde{\Psi}^{\Box}$ and full polynomial exactness of order m and $\tilde{m}$, respectively. Such a basis $\Psi^{\Box}$ can be derived by a d–fold tensor product of the corresponding interval bases introduced in the previous subsection. In order to obtain a convenient notation, we shall use the wavelet index sets

$$\begin{aligned} \mathcal{J}^{\Box} :=& \big\{ (j_0, \mathbf{0}, \mathbf{k}) : k_i \in \Delta_j, 1 \leq i \leq d \big\} \\ & \cup \big\{ (j, \mathbf{e}, \mathbf{k}) : j \geq j_0, \mathbf{0} \neq \mathbf{e} \in \{0,1\}^d, k_i \in \nabla_{j,e_i}, 1 \leq i \leq d \big\}, \end{aligned} \tag{1.3.29}$$

where $\nabla_{j,e}$ are the admissible translation parameters k for interval generators or wavelets on the level j

$$\nabla_{j,e} := \begin{cases} \Delta_j, & e = 0 \\ \nabla_j, & e = 1 \end{cases}. \tag{1.3.30}$$

Then the tensor product wavelets are simply given by

$$\psi_\lambda^{\Box}(x) := \prod_{i=1}^{d} \psi_{(j,e_i,k_i)}(x_i), \quad x \in \Box, \quad \lambda = (j, \mathbf{e}, \mathbf{k}) \in \mathcal{J}^{\Box}. \tag{1.3.31}$$

For the construction of a global composite wavelet basis Ψ over Ω as introduced in [62], the principal design pattern of stable completion can be used again. In a first step, a globally continuous biorthogonal system of generators is constructed that span global multiresolution spaces $S_j, \tilde{S}_j \subset L_2(\Omega)$. This can be done by glueing together the local primal and dual generator bases appropriately across the lower–dimensional interfaces between adjacent subpatches, see [9, 62, 102] for the technical details. As a consequence of this construction, the global Sobolev regularity of the primal and dual composite generators is limited by $\frac{3}{2}$, whereas their patchwise regularity may be significantly higher. It must be noted that the glueing procedure requires further properties from the interval wavelet bases. More precisely, exactly one primal and one dual interval generator is allowed to have a nonzero point value at $x \in \{0,1\}$, respectively. This assumption is fulfilled by special variants of the constructions [59, 62], but unfortunately not for any of the spline wavelet bases from

[126]. There, more than one of the dual generators on the level j does not vanish at a given endpoint of the interval. Consequently, we will use the wavelet bases from [59, 62] for the numerical experiments on the L–shaped domain, where we employ the stabilization techniques from [9] already mentioned in the last subsection.

An example of a primal composite spline generator on the L–shaped domain $\Omega = (-1, 1)^2 \setminus [0, 1)^2$, based in an interval basis from [126], can be found in Figure 1.5.

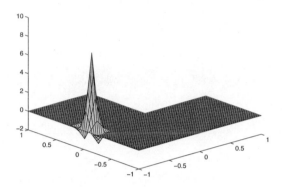

Figure 1.5: A primal composite generator on the L–shaped domain with level $|\lambda| = 3$, belonging to the interface between two adjacent subpatches. The interval basis is taken from [9, 62] with parameters $m = \tilde{m} = 2$.

Given the biorthogonal system of generators, an initial stable completion for some complement spaces W_j can then be constructed from the lifted local wavelet bases, and a final biorthogonalization step of the form (1.3.20) yields the global wavelet system Ψ. For the intricate technical details, we refer the reader to [62]. Let us only remark that the global wavelets ψ_λ have a local decomposition in the lifted auxiliary bases

$$\psi_\lambda|_{\Omega_i}(x) = \sum_{\mu \in \mathcal{I}(\lambda, i)} c_{\mu,i} \psi_\mu^{\square}(\kappa_i^{-1}(x)), \quad x \in \Omega_i, \tag{1.3.32}$$

where $\mathcal{I}(\lambda, i)$ is a suitable set of indices from $\mathcal{J}^{\square}$. Examples of primal composite wavelets on the L–shaped domain can be found in Figure 1.6.

Despite the fact that the dual composite wavelets are also globally continuous, the lower bound $-\tilde{\gamma}$ in the range of Sobolev spaces that can be characterized with a given composite wavelet basis is larger than $-\frac{1}{2}$. This is due to the fact that the construction in [62] uses a modified L_2–inner product that differs from the natural one. However, for the numerical treatment of second order differential equations, this limited range is not a principal issue.

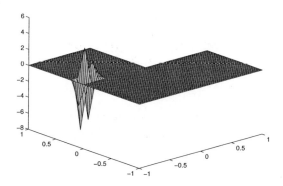

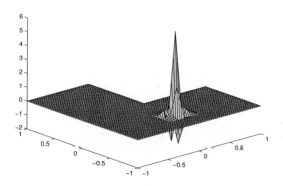

Figure 1.6: Primal composite wavelets ψ_λ on the L–shaped domain with $|\lambda| = 3$ and $m = \tilde{m} = 2$, using the interval basis from [9, 62]. The upper wavelet is of the type $\mathbf{e} = (1,0)$ and it belongs to the subpatch $\Omega_1 = (-1,0) \times (0,1)$. The lower wavelet is of the type $\mathbf{e} = (0,1)$ and it belongs to the interface $\{0\} \times (-1,0)$ between two adjacent subpatches.

Chapter 2

Frames

This chapter is concerned with the definition and the construction of frames on a bounded domain Ω in $\mathbb{R}^d$. We are particularly interested in those classes of frames that are suitable for the numerical discretization of elliptic operator equations. Unless otherwise specified, the results from this chapter can be found in [51].

In Section 2.1, we review the classical definition and properties of frames in a Hilbert space. It is shown how to reformulate the concept of wavelet Riesz bases with norm equivalences in a frame setting, resulting in the definition of Gelfand frames. Section 2.2 is devoted to the construction of specific wavelet Gelfand frames on a polygonal domain that may be used in the numerical treatment of second order elliptic operator equations. For the verification of the Gelfand frame property, we shall exploit the machinery of localized frames to a considerable extent.

2.1 Hilbert and Gelfand Frames

In the following, we will introduce the concept of classical Hilbert as well as Gelfand frames in a separable Hilbert space V. The corresponding inner product shall be denoted by $\langle \cdot, \cdot \rangle_V$, with induced norm $\|v\|_V := \langle v, v \rangle_V^{1/2}$.

2.1.1 Basic Frame Theory

Let us assume that $\mathcal{J}$ is a countable index set. As usual, a system $\Psi = \{\psi_\lambda\}_{\lambda \in \mathcal{J}} \subset V$ is called a *Hilbert frame* or simply a *frame* for V, if there exist *frame bounds* $A, B > 0$ such that the following condition holds:

$$A\|f\|_V^2 \leq \sum_{\lambda \in \mathcal{J}} |\langle f, \psi_\lambda \rangle_V|^2 \leq B\|f\|_V^2, \quad f \in V. \tag{2.1.1}$$

In the sequel, let us recall some further fundamental properties of frames that immediately follow from their very definition. The corresponding proofs can be found in the basic textbook [29].

First of all, it can be shown that any Riesz basis Ψ for V with Riesz constants c_V, C_V is also a frame for V, with frame constants $A = c_V^2$ and $B = C_V^2$, see [29]. Conversely, any frame Ψ for V is automatically dense, as it is the case for Riesz bases. However, the frame condition (2.1.1) does not imply that the elements of Ψ are linearly independent. Thus the coefficient array of an element $f \in \text{clos}_V \text{span} \, \Psi$ may not necessarily be unique, rendering the concept of frames a proper generalization of the Riesz basis concept.

Any frame Ψ with frame bounds A, B readily induces several fundamental operators. Firstly the *frame operator*

$$S : V \to V, \quad f \mapsto \sum_{\lambda \in \mathcal{J}} \langle f, \psi_\lambda \rangle_V \psi_\lambda \tag{2.1.2}$$

is well–defined and bounded with $\|S\|_{\mathcal{L}(V)} \leq B$, and the series Sf converges unconditionally. Moreover, S is self–adjoint, positive and boundedly invertible with

$$A \langle f, f \rangle_V \leq \langle Sf, f \rangle_V \leq B \langle f, f \rangle_V, \quad f \in V. \tag{2.1.3}$$

S can be written as a composition of the *analysis operator*

$$F : V \to \ell_2, \quad Ff := \langle f, \Psi \rangle_V^\top \tag{2.1.4}$$

and its ℓ_2–adjoint

$$F^* : \ell_2 \to V, \quad F^* \mathbf{c} := \mathbf{c}^\top \Psi, \tag{2.1.5}$$

the *synthesis operator*. F is a bounded injective mapping with $\|F\|_{\mathcal{L}(\ell_2, V)} \leq \sqrt{B}$ and closed range $\text{Ran}(F) = \text{clos}_V \text{Ran}(F)$, whereas F^* is bounded and onto. In fact, the latter two conditions on F^* are equivalent to Ψ being a frame.

As a consequence of the orthogonal decomposition

$$\ell_2 = \text{Ran}(F) \oplus \text{Ker}(F^*), \tag{2.1.6}$$

a given frame Ψ for V is a Riesz basis if and only if $\text{Ran}(F) = V$, which in turn is equivalent to $\text{Ker} \, F^* = \{0\}$. The orthogonal projection of ℓ_2 onto $\text{Ran} \, F$ is given by the operator

$$\mathbf{Q} := F(F^* F)^{-1} F^* : \ell_2 \to \ell_2. \tag{2.1.7}$$

The system $\tilde{\Psi} = \{\tilde{\psi}_\lambda\}_{\lambda \in \mathcal{J}}$ with $\tilde{\psi}_\lambda := S^{-1} \psi_\lambda$ is again a frame for V, with frame bounds B^{-1} and A^{-1}. $\tilde{\Psi}$ is also called the *canonical dual frame*. The analysis and synthesis operators of $\tilde{\Psi}$ are given by

$$\tilde{F} = F(F^* F)^{-1} = F S^{-1}, \quad \tilde{F}^* = (F^* F)^{-1} F^* = S^{-1} F^*, \tag{2.1.8}$$

and $S^{-1} : V \to V$ is the frame operator for $\tilde{\Psi}$. From the bounded invertibility of S we infer that any $f \in V$ can be reconstructed by the unconditionally convergent expansions

$$f = S^{-1} S f = \sum_{\lambda \in \mathcal{J}} \langle f, \psi_\lambda \rangle_V S^{-1} \psi_\lambda = S S^{-1} f = \sum_{\lambda \in \mathcal{J}} \langle f, S^{-1} \psi_\lambda \rangle_V \psi_\lambda. \tag{2.1.9}$$

For a frame Ψ, in the nontrivial case of $\text{Ker} \, F^* \neq \{0\}$, the coefficient functionals $\{c_\lambda\}_{\lambda \in \mathcal{J}} \subset V'$ in a representation

$$f = \sum_{\lambda \in \mathcal{J}} c_\lambda(f) \psi_\lambda, \quad f \in V, \tag{2.1.10}$$

may not be unique. Hence there may well exist other *non–canonical* dual frames $\Xi = \{\xi_\lambda\}_{\lambda \in \mathcal{J}}$ which also fulfill a reconstruction equation of the form

$$f = \sum_{\lambda \in \mathcal{J}} \langle f, \psi_\lambda \rangle_V \xi_\lambda = \sum_{\lambda \in \mathcal{J}} \langle f, \xi_\lambda \rangle_V \psi_\lambda, \quad f \in V. \tag{2.1.11}$$

However then, the analogous operator to $\mathbf{Q}$ from (2.1.7)

$$\mathbf{P}_\Xi : \ell_2 \to \ell_2, \quad \mathbf{P}_\Xi \mathbf{c} := \Big(\sum_{\mu \in \mathcal{J}} \langle \mathbf{c}^\top \Psi, \xi_\mu \rangle \langle \xi_\mu, \psi_\lambda \rangle \Big)_{\lambda \in \mathcal{J}} \tag{2.1.12}$$

is only a projector onto Ran F and not orthogonal.

2.1.2 Gelfand Frames

As we have seen in Chapter 1, wavelet Riesz bases Ψ are almost always constructed in $L_2(\Omega)$ first. Then in a second step, utilizing approximation and smoothness properties of the corresponding approximating spaces S_j and $\tilde{S}_j$, it is shown that appropriately rescaled versions of the system Ψ are also Riesz bases in other smoothness spaces.

It is now our aim to generalize this very strategy towards the case of frames. Namely, we shall introduce a special class of frames in a Hilbert space V that, when rescaling the frame elements appropriately, gives rise to a frame for another Hilbert space H. We will have to investigate whether the expansion (2.1.9) also converges in other topologies than in that of V.

To this end, we shall recall first a powerful concept from [86, 89]. Given a separable and reflexive Banach space B with normed dual B', a system $\mathcal{G} = \{g_\theta\}_{\theta \in \Theta} \subset B'$ with associated Banach sequence space b is called a *Banach frame* for B, if the coefficient operator

$$G : B \to b, \quad f \mapsto \langle f, \mathcal{G} \rangle_{B \times B'}^\top \tag{2.1.13}$$

is bounded with the norm equivalence

$$\|Gf\|_b \approx \|f\|_B, \quad f \in B, \tag{2.1.14}$$

and there exists a bounded left–inverse $R : b \to B$ of G, the so–called *reconstruction operator*. In fact, the concept of Banach frames generalizes that of Hilbert frames, as is easily shown in the following lemma.

Lemma 2.1. *Any frame* $\Psi = \{\psi_\lambda\}_{\lambda \in \mathcal{J}}$ *for a separable Hilbert space* V *is also a Banach frame for* V, *with associated sequence space* $\ell_2 = \ell_2(\mathcal{J})$.

Proof. Given a frame Ψ for V, let $\mathcal{G} := \tilde{\Psi} \subset V$ be the canonical dual frame. Identifying V and V' via the Riesz map, we can assume that $\mathcal{G} = \{\langle \cdot, \tilde{\psi}_\lambda \rangle_V\}_{\lambda \in \mathcal{J}} \subset V'$. The operator G from (2.1.13) then turns out to be the dual analysis operator $\tilde{F}$ from (2.1.8), so that we can choose the reconstruction operator $R = F^*$, which is a left–inverse of $\tilde{F}$. $\square$

For the discretization of elliptic operator equations, we will have to work with expansions both in the Hilbert space V and in a densely embedded Banach space $B \hookrightarrow V$. More precisely, we shall work in a Gelfand triple situation (B, V, B') where

$$B \subset V \simeq V' \subset B', \tag{2.1.15}$$

and we assume explicitly that the right inclusion is also dense. In particular, this holds if B is also a Hilbert space.

A special class of frames tailored to this Gelfand frame situation has been introduced in [51]. Namely, a given frame Ψ for V with canonical dual frame $\tilde{\Psi}$ is called a *Gelfand frame* for the Gelfand triple (B, V, B'), if $\Psi \subset B$, $\tilde{\Psi} \subset B'$ and there exists a Gelfand triple $(b, \ell_2(\mathcal{J}), b')$ of sequence spaces such that

$$F^* : b \to B, \ F^*\mathbf{c} = \mathbf{c}^\top \Psi \quad \text{and} \quad \tilde{F} : B \to b, \ \tilde{F}f = \big(\langle f, \tilde{\psi}_\lambda \rangle_{B \times B'} \big)_{\lambda \in \mathcal{J}} \tag{2.1.16}$$

are bounded operators. By duality arguments in the Gelfand triple (B, V, B'), we can infer from the identity $\tilde{F}^* F = F^* \tilde{F} = \mathrm{id}_V$ that for a Gelfand frame Ψ, also the operators

$$\tilde{F}^* : b' \to B', \ \tilde{F}^*\mathbf{c} = \mathbf{c}^\top \tilde{\Psi} \quad \text{and} \quad F : B' \to b', \ Ff = \big(\langle f, \psi_\lambda \rangle_{B' \times B} \big)_{\lambda \in \mathcal{J}} \tag{2.1.17}$$

are bounded, cf. [92]. For a visualization of the various mappings in a Gelfand frame, we refer the reader to Figure 2.1.

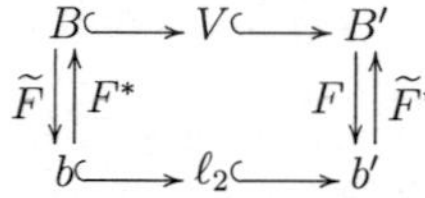

$$
\begin{array}{ccc}
B \hookrightarrow & V \hookrightarrow & B' \\
\tilde{F} \big\uparrow\big\downarrow F^* & & F \big\downarrow\big\uparrow \tilde{F}^* \\
b \hookrightarrow & \ell_2 \hookrightarrow & b'
\end{array}
$$

Figure 2.1: Mapping diagram for a Gelfand frame Ψ

The next result clarifies the relations between Gelfand and Banach frames:

Proposition 2.2. *If Ψ is a Gelfand frame for (B, V, B') with the Gelfand triple of sequence spaces $(b, \ell_2(\mathcal{J}), b')$, then $\tilde{\Psi}$ and Ψ are Banach frames for B and B', respectively.*

Proof. We only show that $\tilde{\Psi}$ is a Banach frame for B, since the second claim follows by an analogous argument. By definition of a Gelfand frame, it is $\tilde{\Psi} \subset B'$, and the coefficient operator $G : B \to b$ is given by $Gf = \tilde{F}f = \langle f, \tilde{\Psi} \rangle_{B \times B'}^\top$. Since $\tilde{\Psi}$ is the canonical dual of Ψ, we have by (2.1.9) the representation $f = \langle f, \tilde{\Psi} \rangle_V \Psi$ for each $f \in V$, with convergence in V. But for $f \in B$, we have $\langle f, \tilde{\psi}_\lambda \rangle_V = \langle f, \tilde{\psi}_\lambda \rangle_{B \times B'}$ by the definition of the dual pairing. Since, moreover, $F^* \tilde{F} f \in B$ by the boundedness of F^* and $\tilde{F}$, the series $\sum_{\lambda \in \mathcal{J}} \langle f, \tilde{\psi}_\lambda \rangle_{B \times B'} \psi_\lambda = F^* \tilde{F} f = f$ also converges in B. Finally, this yields the injectivity of $G = \tilde{F}$

$$\|f\|_B = \Big\| \sum_{\lambda \in \mathcal{J}} \langle f, \tilde{\psi}_\lambda \rangle_{B \times B'} \psi_\lambda \Big\|_B = \|F^* \tilde{F} f\|_B \lesssim \|\tilde{F} f\|_b \lesssim \|f\|_B.$$

A bounded reconstruction operator R is of course given by $F^* : b \to B$. $\qquad\square$

In the original definition of Gelfand frames, the sequence spaces b and ℓ_2 are completely unrelated, up to the dense and continuous embedding $b \hookrightarrow \ell_2$. This is mainly due to the fact that b is in general an arbitrary Banach sequence space and therefore structurally different from the Hilbert space ℓ_2. However, in applications we may assume that b has more stringent structural properties. As an example, when it comes to frame discretizations of elliptic differential operators as in Chapter 5, the role of b will be played by a weighted ℓ_2 space. In view of these situations, for any strictly positive diagonal matrix $\mathbf{W} = \mathrm{diag}(w_\lambda)_{\lambda \in \mathcal{J}} > 0$ and $1 < p < \infty$, let us introduce the weighted ℓ_p spaces

$$\ell_{p,\mathbf{W}} := \{\mathbf{c} : \|\mathbf{c}\|_{\ell_{p,\mathbf{W}}} := \|\mathbf{W}\mathbf{c}\|_{\ell_p} < \infty\}. \tag{2.1.18}$$

Then we can immediately relate $\ell_{2,\mathbf{W}}$ and ℓ_2 by an isomorphism

$$\varphi_{\mathbf{W}} : \ell_{2,\mathbf{W}} \to \ell_2, \quad \varphi_{\mathbf{W}}\mathbf{c} = \mathbf{W}\mathbf{c} \tag{2.1.19}$$

with structurally identical dual mapping

$$\varphi_{\mathbf{W}}^* : \ell_2 \to \ell_{2,\mathbf{W}^{-1}}, \quad \varphi_{\mathbf{W}}^*\mathbf{c} = \mathbf{W}\mathbf{c}. \tag{2.1.20}$$

Both mappings $\varphi_{\mathbf{W}}$ and $\varphi_{\mathbf{W}}^*$ have operator norm 1 in the respective topologies. Given that the abstract sequence space b in the Gelfand frame definition is in fact such a weighted ℓ_2 space, we can strengthen Proposition 2.2 substantially in the following way:

Proposition 2.3. *Let H be a Hilbert space and $\Psi = \{\psi_\lambda\}_{\lambda \in \mathcal{J}}$ be a Gelfand frame for (H, V, H') with the Gelfand triple of sequence spaces $(\ell_{2,\mathbf{W}}, \ell_2, \ell_{2,\mathbf{W}^{-1}})$, where $w : \mathcal{J} \to \mathbb{R}_+$ is a strictly positive weight function. Then the systems $\mathbf{W}^{-1}\Psi = \{w_\lambda^{-1}\psi_\lambda\}_{\lambda \in \mathcal{J}}$ and $\mathbf{W}\tilde{\Psi} = \{w_\lambda\tilde{\psi}_\lambda\}_{\lambda \in \mathcal{J}}$ are (Hilbert) frames for H and H', respectively.*

Proof. We only prove that $\mathbf{W}^{-1}\Psi$ is a Hilbert frame in H, since the other claim follows by analogy.

Firstly, Ψ being a Gelfand frame for (H, V, H'), we know that the operator $F^* : \ell_{2,\mathbf{W}} \to H$, $F^*\mathbf{c} = \mathbf{c}^\top\Psi$, is bounded. The composition $T := F^*\varphi_{\mathbf{W}}^{-1} : \ell_2 \to H$, being the synthesis operator of the system $\mathbf{W}^{-1}\Psi \subset H$, is also bounded, so that $\mathbf{W}^{-1}\Psi$ is a Bessel system in H which is equivalent to the validity of the upper frame bound, see also [29].

Concerning the lower frame bound, we utilize from Proposition 2.2 that $(\tilde{\Psi}, \ell_{2,\mathbf{W}})$ is a Banach frame for H, with bounded reconstruction operator $R = F^* : \ell_{2,\mathbf{W}} \to H$. Since F^* is onto, this is also the case for the bounded operator $T := F^*\varphi_{\mathbf{W}}^{-1} : \ell_2 \to H$ already considered above. Hence, for the bounded pseudoinverse $T^\dagger = \varphi_{\mathbf{W}}\tilde{F} : H \to \ell_2$, we have $f = TT^\dagger f$, so that the lower frame bound follows from

$$\begin{aligned}
\|f\|_H^4 &= \left|\langle TT^\dagger f, f\rangle_H\right|^2 \\
&= \left|\sum_{\lambda \in \mathcal{J}} \langle (T^\dagger f)_\lambda w_\lambda^{-1}\psi_\lambda, f\rangle_H\right|^2 \\
&\leq \sum_{\mu \in \mathcal{J}} |(T^\dagger f)_\mu|^2 \sum_{\lambda \in \mathcal{J}} |\langle f, w_\lambda^{-1}\psi_\lambda\rangle_H|^2 \\
&\leq \|T^\dagger\|_{\mathcal{L}(H,\ell_2)}^2 \|f\|_H^2 \sum_{\lambda \in \mathcal{J}} |\langle f, w_\lambda^{-1}\psi_\lambda\rangle_H|^2.
\end{aligned}$$

$\square$

Remark 2.4. *Proposition 2.3 also sheds some light on the slightly different settings in* [51, 133]. *In both papers, frames are considered for the numerical discretization of an elliptic operator equation* $Au = f$ *with an isomorphism* $A : H \to H'$.

In [133], *a frame* Ψ *for* H *is used without explicitly requiring that a rescaled version of* Ψ *is also an* L_2-*frame. Moreover, the energy space* H *is implicitly identified with its normed dual* H' *via the Riesz mapping. For the numerical approximation of* u *in* H, *this setting is sufficient since a frame* Ψ *for* H *is automatically* H-*dense. The assumptions then cover also the case where the underlying frame is* not *of wavelet type, cf.* [123, 124].

Conversely, the more restricted setting of Gelfand frames in [51], *as introduced in this section, conforms to the well-known constructions of wavelet Riesz bases. The results in loc. cit. make use of localization arguments that require the canonical dual frame* $\tilde{\Psi}$ *to be contained and to be stable in* L_2, *see Section 2.2. Moreover, it is assumed there to have a Gelfand triple setting where the middle space* L_2 *is identified with its normed dual.*

2.2   Aggregated Gelfand Frames

2.2.1   General Idea

Assume that the domain $\Omega \subset \mathbb{R}^d$ can be written as an overlapping union of M patches Ω_i, $1 \le i \le M$

$$\Omega = \bigcup_{i=1}^{M} \Omega_i, \tag{2.2.1}$$

where each subpatch is the image of the unit cube $\square := (0,1)^d$ under a suitable parametrization $\Omega_i = \kappa_i(\square)$. We assume that the parametrizations κ_i are C^k-diffeomorphisms and that

$$|\det D\kappa_i(\mathbf{x})| \approx 1, \quad \mathbf{x} \in \square. \tag{2.2.2}$$

We can hence assume that the locality estimate

$$\#\{\mathbf{x} \in \square : \kappa_i(\mathbf{x}) = \mathbf{y} \text{ for some } i\} \le M \tag{2.2.3}$$

holds uniformly in $\mathbf{y} \in \Omega$. Clearly, the set of admissible domains Ω is restricted by raising these regularity conditions. Essentially, the boundary of Ω has to be piecewise smooth enough. But since the particularly attractive case of polyhedral domains is still covered, these assumptions on the parametrizations κ_i are valid for the boundary value problems under consideration.

Let us assume furthermore that we have a reference frame $\Psi^\square = \{\psi_\mu\}_{\mu \in \mathcal{J}^\square} \subset L_2(\square)$. Then, one can lift the system $\Psi^\square$ to Ω_i by setting

$$\psi_{i,\mu} := \frac{\psi_\mu^\square(\kappa_i^{-1}(\cdot))}{|\det D\kappa_i(\kappa_i^{-1}(\cdot))|^{1/2}}. \tag{2.2.4}$$

Note that the denominator is chosen in such a way that $\|\psi_{i,\mu}\|_{L_2(\Omega)} = \|\psi_\mu^\square\|_{L_2(\square)}$. Analogously, we also lift the dual frame elements to Ω_i:

$$\tilde{\psi}_{i,\mu} := \frac{\tilde{\psi}_\mu^\square\big(\kappa_i^{-1}(\cdot)\big)}{\big|\det D\kappa_i\big(\kappa_i^{-1}(\cdot)\big)\big|^{1/2}}. \tag{2.2.5}$$

It is immediate to see that both the Hilbert and the Gelfand frame properties transfer to the lifted systems on Ω_i:

Lemma 2.5. *Let $\Psi^\square$ be a Gelfand frame in $(H_0^s(\square), L_2(\square), H^{-s}(\square))$ for the Gelfand triple of sequence spaces $(\ell_{2,\mathbf{D}^s}(\mathcal{J}^\square), \ell_2(\mathcal{J}^\square), \ell_{2,\mathbf{D}^{-s}}(\mathcal{J}^\square))$. Then the system $\Psi^{(i)} := \{\psi_{i,\mu} : \mu \in \mathcal{J}^\square\}$ as defined in (2.2.4) is a Gelfand frame in $(H_0^s(\Omega_i), L_2(\Omega_i), H^{-s}(\Omega_i))$ for the Gelfand triple of sequence spaces $(\ell_{2,\mathbf{D}^s}(\mathcal{J}^\square), \ell_2(\mathcal{J}^\square), \ell_{2,\mathbf{D}^{-s}}(\mathcal{J}^\square))$. Moreover, the canonical dual frame elements of $\Psi^{(i)}$ are exactly the lifted reference duals $\tilde{\Psi}^{(i)} := \{\tilde{\psi}_\lambda\}_{\lambda \in \mathcal{J}^\square}$ from (2.2.5).*

Proof. For $f \in L_2(\Omega_i)$, (2.2.2) and a transformation of coordinates imply

$$\|f\|_{L_2(\Omega_i)}^2 \eqsim \big\||f(\cdot)|\det D\kappa_i(\kappa_i^{-1}(\cdot))|^{1/2}\big\|_{L_2(\Omega_i)}^2 \eqsim \big\||f \circ \kappa_i(\cdot)|\det D\kappa_i(\cdot)|^{1/2}\big\|_{L_2(\square)}^2.$$

Inserting the frame condition for $\Psi^\square$ in $L_2(\square)$, we get the frame condition for $\Psi^{(i)}$ in $L_2(\Omega_i)$

$$\|f\|_{L_2(\Omega_i)}^2 \eqsim \sum_{\mu \in \mathcal{J}^\square}\big|\langle f \circ \kappa_i(\cdot)|\det D\kappa_i(\cdot)|^{1/2}, \psi_\mu^\square\rangle_{L_2(\square)}\big|^2 = \sum_{\mu \in \mathcal{J}^\square}\big|\langle f, \psi_{i,\mu}\rangle_{L_2(\Omega_i)}\big|^2. \tag{2.2.6}$$

Let $S^\square$ and $S^{(i)}$ be the frame operators for $\Psi^\square$ and $\Psi^{(i)}$, respectively. Then for $\mu \in \mathcal{J}^\square$ and $y = \kappa_i(x) \in \Omega_i$, it follows that

$$\begin{aligned}
S^{(i)}\tilde{\psi}_{i,\mu}(y) &= \sum_{\nu \in \mathcal{J}^\square}\langle \tilde{\psi}_{i,\mu}, \psi_{i,\nu}\rangle_{L_2(\Omega_i)}\psi_{i,\nu}(y) \\
&= |\det D\kappa_i(x)|^{-1/2}\sum_{\nu \in \mathcal{J}^\square}\langle \tilde{\psi}_\mu^\square, \psi_\nu^\square\rangle_{L_2(\square)}\psi_\nu^\square(x) \\
&= |\det D\kappa_i(x)|^{-1/2}S^\square\tilde{\psi}_\mu^\square(x) \\
&= |\det D\kappa_i(x)|^{-1/2}\psi_\mu^\square(x) \\
&= \psi_{i,\mu}(y),
\end{aligned}$$

so that $\tilde{\psi}_{i,\mu}$ is the canonical dual frame element corresponding to $\psi_{i,\mu}$. For a sequence $\mathbf{c} \in \ell_{2,\mathbf{D}^s}(\mathcal{J}^\square)$, it follows by (2.2.2), the chain rule for Sobolev norms and (2.2.4) that

$$\Big\|\sum_{\mu \in \mathcal{J}^\square}c_\mu\psi_{i,\mu}\Big\|_{H^s(\Omega_i)} \lesssim \Big\|\sum_{\mu \in \mathcal{J}^\square}c_\mu\psi_{i,\mu} \circ \kappa_i\Big\|_{H^s(\square)} = \Big\||\det D\kappa_i(\cdot)|^{-1/2}\sum_{\mu \in \mathcal{J}^\square}c_\mu\psi_\mu^\square\Big\|_{H^s(\square)}.$$

Hence an application of the product rule for Sobolev norms gives

$$\Big\|\sum_{\mu \in \mathcal{J}^\square}c_\mu\psi_{i,\mu}\Big\|_{H^s(\Omega_i)} \lesssim \Big\|\sum_{\mu \in \mathcal{J}^\square}c_\mu\psi_\mu^\square\Big\|_{H^s(\square)} \lesssim \|\mathbf{c}\|_{\ell_{2,\mathbf{D}^s}(\mathcal{J}^\square)},$$

so the operator $(F^{(i)})^* : \ell_{2,\mathbf{D}^s}(\mathcal{J}^\square) \to H_0^s(\Omega_i)$, $(F^{(i)})^* \mathbf{c} = \mathbf{c}^\top \Psi^{(i)}$ is bounded. Moreover, knowing the canonical dual elements of $\Psi^{(i)}$, it is for an $f \in H_0^s(\Omega_i)$

$$\sum_{\mu \in \mathcal{J}^\square} |\langle f, \tilde{\psi}_{i,\mu}\rangle_{H_0^s(\Omega_i) \times H^{-s}(\Omega_i)}|^2 = \sum_{\mu \in \mathcal{J}^\square} \left| \langle f \circ \kappa_i(\cdot) | \det D\kappa_i(\cdot)|^{1/2}, \tilde{\psi}_\mu^\square \rangle_{H_0^s(\square) \times H^{-s}(\square)} \right|^2$$

$$\lesssim \left\| f \circ \kappa_i(\cdot)| \det D\kappa_i(\cdot)|^{1/2} \right\|_{H^s(\square)}$$

$$\lesssim \| f \circ \kappa_i \|_{H^s(\square)}$$

$$\lesssim \| f \|_{H^s(\Omega_i)},$$

so the operator $\widetilde{F}^{(i)} : H_0^s(\Omega_i) \to \ell_{2,\mathbf{D}^s}(\mathcal{J}^\square)$, $\widetilde{F}^{(i)} f = (\langle f, \tilde{\psi}_{i,\mu}\rangle_{H_0^s(\Omega_i) \times H^{-s}(\Omega_i)})_{\mu \in \mathcal{J}^\square}$ is bounded, which yields the claim. $\square$

Remark 2.6. *The reader might ask whether we may also omit the denominator in the definition* (2.2.4) *of* $\psi_{i,\mu}$. *As becomes visible in the proof of Lemma 2.5, this variant still gives an* L_2 *frame and it is maybe more convenient from the computational point of view, as soon as a nontrivial chart is involved. However, the norms of* $\|\psi_{i,\mu}\|_{L_2(\Omega_i)}$ *will in general be different from* $\|\psi_\mu^\square\|_{L_2(\square)}$. *Moreover, the lifted dual reference frame* $\tilde{\Psi}^{(i)}$ *then no longer coincides with the canonical dual of the lifted primal reference frame* $\Psi^{(i)}$.

Our goal is now to construct a Gelfand frame Ψ over the domain Ω, by aggregating the local Gelfand frames in the most simple way. To this end, we define the index set

$$\mathcal{J} := \bigcup_{i=1}^M \{i\} \times \mathcal{J}^\square, \tag{2.2.7}$$

and, for each $\lambda = (i, \mu) \in \mathcal{J}$, we set $|\lambda| := |\mu|$. In the following, we will consider the aggregated system

$$\Psi := \{\psi_\lambda\}_{\lambda \in \mathcal{J}}, \quad \psi_{(i,\mu)} := E_i \psi_{i,\mu}, \text{ for } (i, \mu) \in \mathcal{J}, \tag{2.2.8}$$

where, for each subpatch Ω_i, $E_i : \Omega_i \to \Omega$ denotes the extension by zero. Note that, by definition of the Sobolev spaces with homogeneous boundary conditions, the operators E_i are bounded with norm 1 from $H_0^s(\Omega_i)$ to $H_0^s(\Omega)$ for any $s \geq 0$.

First of all, it is immediate to see the Hilbert frame property of Ψ.

Lemma 2.7. *The system* Ψ *is a Hilbert frame in* $L_2(\Omega)$.

Proof. For $f \in L_2(\Omega)$, the frame condition for Ψ follows by summing (2.2.6) up over i and using the trivial equivalence

$$\|f\|_{L_2(\Omega)}^2 \leq \sum_{i=1}^M \|f\|_{L_2(\Omega_i)}^2 \leq M \|f\|_{L_2(\Omega)}^2.$$

$\square$

Remark 2.8. *In* [133], *it was shown that the system* Ψ *is a Hilbert frame in* $H_0^s(\Omega)$, *using a specific partition of unity subordinate to the overlapping covering of* Ω. *However, not all domains of the form* (2.2.1) *admit the partition of unity arguments in* [133], *especially not the L–shaped domain. Therefore, alternative proofs for the Hilbert and Gelfand frame properties of* Ψ *are necessary.*

The next goal would be, of course, to clarify under which conditions the frame Ψ can also be a Gelfand frame in $(\Pi_0^s(\Omega), L_2(\Omega), H^{-s}(\Omega))$. However, as one has to check the boundedness of the operator

$$\widetilde{F} : H_0^s(\Omega) \rightarrow \ell_{2,\mathbf{D}^s}(\mathcal{J}), \quad f \mapsto \left(\langle f, \widetilde{\psi_\lambda}\rangle\right)_{\lambda \in \mathcal{J}} \tag{2.2.9}$$

it turns out, here one needs more information about the *global* canonical dual $\widetilde{\Psi} = \{\widetilde{\psi_\lambda}\}_{\lambda \in \mathcal{J}}$ of Ψ. Our main tool for gaining knowledge about the canonical dual frame will be based on the theory of localized frames which shall be sketched in the next subsection.

2.2.2 Localization of Frames

To ensure that the system Ψ is a Gelfand frame, there are several possible sufficient conditions. One of them is based on the concept of localized frames. Generally speaking, two frames $\mathcal{F} = \{f_\theta\}_{\theta \in \Theta}$ and $\mathcal{G} = \{g_{\theta'}\}_{\theta' \in \Theta}$ for a Hilbert space V are called *localized* to each other, if the entries $\langle f_\theta, g_{\theta'}\rangle_V$ of their cross–Gramian matrix exhibit some decay in the term $\rho(\theta, \theta')$, where

$$\rho : \Theta \times \Theta \rightarrow \mathbb{R}_+^0$$

is an appropriate distance function on the ordered index set Θ. In the original definition of localized frames in [88, 89], the special case $\Theta = \mathbb{Z}$ and the metric $\rho(\theta, \theta') = |\theta - \theta'|$ were considered. However, many of the results and ideas readily transfer also to more complicated index sets and more general distance functions. The essential properties of ρ shall be discussed in the following.

Typically, localized frames exhibit either polynomial or exponential decay estimates in the index distance $\rho(\theta, \theta')$. Here we are particularly interested in decay estimates of *Lemarié type* where the Gramian matrix is in the *Lemarié class*. Under the assumption that the index set Θ admits a scale mapping of the form $\Theta \ni \theta \mapsto |\theta| \in \mathbb{Z}_{\geq j_0}$, then, for fixed parameters $\beta, \sigma > 0$ and a function $\rho : \Theta \times \Theta \rightarrow \mathbb{R}_+^0$, the *Lemarié class* $\mathcal{A}_{\sigma,\beta} = \mathcal{A}_{\sigma,\beta,\rho}$ is the set of all matrices $\mathbf{B} = (b_{\theta,\theta'})_{\theta,\theta' \in \Theta}$, such that

$$|b_{\theta,\theta'}| \leq c_{\mathbf{B}} 2^{-\||\theta| - |\theta'|\|}(1 + \rho(\theta,\theta'))^{-\beta}, \quad \theta, \theta' \in \mathcal{J} \tag{2.2.10}$$

holds for a constant $c_{\mathbf{B}}$ which only depends on $\mathbf{B}$. It is of particular importance that the right–hand side in (2.2.10) can be transformed into a special exponential term with respect to another distance function $\varrho : \Theta \times \Theta \rightarrow \mathbb{R}_+^0$

$$2^{-\||\theta| - |\xi|\|\sigma}\left(1 + \rho(\theta, \xi)\right)^{-\beta} = e^{-\sigma(\||\theta| - |\xi|\| \log 2 + \frac{\beta}{\sigma} \log(1 + \rho(\theta,\xi)))} =: e^{-\sigma\varrho(\theta,\xi)}. \tag{2.2.11}$$

For many cases of interest, ϱ fulfills the axioms of a metric, at least in a generalized sense. Therefore we call ϱ a *Lemarié metric* in the sequel. The validity of a generalized triangle inequality will turn out to be of particular importance when dealing with products of matrices from $\mathcal{A}_{\sigma,\beta}$, see Lemma 2.11 and Theorem 2.12 below for details.

However, it should be noted that there are important examples where ϱ and its generating distance function ρ are *not* a metric. More precisely, as we will see in Chapter 4, the Lemarié class $\mathcal{A}_{\sigma,\beta}$ contains matrices that arise in the wavelet discretization of local and non–local elliptic operators $A : H \to H'$, where the energy space H is a closed subspace of $H^t(\Omega)$ and $\rho = \delta$ is given by

$$\delta(\lambda, \lambda') := 2^{\min\{|\lambda|,|\lambda'|\}} \operatorname{dist}(\operatorname{supp} \psi_\lambda, \operatorname{supp} \psi_{\lambda'}), \quad \lambda, \lambda' \in \mathcal{J}. \tag{2.2.12}$$

But neither δ nor the corresponding Lemarié distance

$$\varrho_\delta(\lambda, \lambda') := 2^{-\||\lambda|-|\lambda'|\|\sigma}\left(1 + \delta(\lambda, \lambda')\right)^{-\beta} \tag{2.2.13}$$

are a metric, since the triangle inequality does not hold.

Unfortunately, for the introduction of an appropriate localization concept for wavelet frames over bounded domains, a slight generalization of the aforementioned setting has to be made. In that case, the underlying wavelet index sets are no longer isomorphic in general, at least not for those indices belonging to the coarsest level. One possibility to cope with this situation is to use more than one Lemarié metric at the same time.

A *generalized Lemarié metric* for the pair of countable index sets (Θ, Ξ) is a tuple of functions

$$\varrho = \left(\varrho_{\Theta_1 \times \Theta_2}\right)_{\Theta_1, \Theta_2 \in \{\Theta, \Xi\}} \tag{2.2.14}$$

where

$$\varrho_{\Theta_1 \times \Theta_2} : \Theta_1 \times \Theta_2 \to \mathbb{R}_+^0, \quad \Theta_1, \Theta_2 \in \{\Theta, \Xi\} \tag{2.2.15}$$

and the following three conditions hold:

$$\varrho_{\Theta_1 \times \Theta_1}(\xi, \xi) \lesssim 1, \quad \Theta_1 \in \{\Theta, \Xi\}, \ \xi \in \Theta_1, \tag{2.2.16a}$$

$$\varrho_{\Theta \times \Xi}(\theta, \xi) = \varrho_{\Xi \times \Theta}(\xi, \theta), \quad \theta \in \Theta, \ \xi \in \Xi, \tag{2.2.16b}$$

$$\varrho_{\Theta_1 \times \Theta_3}(\theta_1, \theta_3) \leq \varrho_{\Theta_1 \times \Theta_2}(\theta_1, \theta_2) + \varrho_{\Theta_2 \times \Theta_3}(\theta_2, \theta_3), \ \Theta_i \in \{\Theta, \Xi\}, \ \theta_i \in \Theta_i. \tag{2.2.16c}$$

Remark 2.9. *In [51, 113], instead of* (2.2.16c) *a generalized triangle inequality with an additional parameter $w_0 > 0$*

$$\varrho_{\Theta_1 \times \Theta_3}(\theta_1, \theta_3) \leq \varrho_{\Theta_1 \times \Theta_2}(\theta_1, \theta_2) + w_0 \varrho_{\Theta_2 \times \Theta_3}(\theta_2, \theta_3) \tag{2.2.17}$$

was used. However, as it will turn out in the course of this section, all the relevant Lemarié metrics will fulfill (2.2.17) *with $w_0 = 1$, so that this generalization seems unnecessary.*

Using a generalized Lemarié metric, it is straightforward to introduce the corresponding localization concept. Given two frames $\mathcal{F} = \{f_\theta\}_{\theta \in \Theta}$ and $\mathcal{G} = \{g_\xi\}_{\xi \in \Xi}$ for the Hilbert space $\mathcal{H}$ and a Lemarié metric ϱ, we say that $\mathcal{F}$ is ϱ-*exponentially localized* with respect to $\mathcal{G}$ (or simply *exponentially localized* once ϱ is fixed) if there exists some $\alpha > 0$, such that the following decay estimate holds:

$$\left|\langle f_\theta, g_\xi \rangle_H\right| \lesssim e^{-\alpha \varrho_{\Theta \times \Xi}(\theta, \xi)} \quad \text{for all } \theta \in \Theta, \ \xi \in \Xi. \tag{2.2.18}$$

In such a case we write $\mathcal{F} \sim_{\exp} \mathcal{G}$. A frame $\mathcal{F}$ such that $\mathcal{F} \sim_{\exp} \mathcal{F}$, is called *intrinsically ϱ-exponentially localized*.

Remark 2.10. *Due to the assumption* (2.2.16b), *the relation* $\sim_{\exp}$ *is symmetric in the sense that the roles of* Θ *and* Ξ *can be interchanged.*

Given a generalized Lemarié metric ϱ for the pair (Θ, Ξ), we define the class of all matrices with a ϱ–exponential decay rate $\alpha > 0$ as

$$\mathcal{A}_\alpha(\Theta, \Xi) := \big\{ \mathbf{M} = (m_{\theta,\xi})_{\theta \in \Theta, \xi \in \Xi} : |m_{\theta,\xi}| \le c_\mathbf{M} e^{-\alpha \varrho_{\Theta \times \Xi}(\theta,\xi)}, \ \theta \in \Theta, \xi \in \Xi \big\}. \quad (2.2.19)$$

The definition of $\mathcal{A}_\alpha(\Theta, \Xi)$ conforms to (2.2.10) and shall be used to analyze the behavior of matrices with Lemarié–like decay under matrix multiplication and inversion. It turns out, that under an additional condition on the matrices

$$\mathbf{E}_\gamma(\Theta, \Xi) := (e^{-\gamma \varrho_{\Theta \times \Xi}(\theta,\xi)})_{\theta \in \Theta, \xi \in \Xi}, \quad \gamma > 0 \quad (2.2.20)$$

which resembles the row sum criterion from the Schur lemma, a multiplication of two matrices with a Lemarié–like off–diagonal decay yields again a matrix with nearly the same decay rate. The following lemma is a slight generalization of a lemma from [113, section 5] and [97, Proposition 1]:

Lemma 2.11. *Let* ϱ *be a generalized Lemarié metric for the pair* (Θ, Ξ). *Furthermore, assume that* $D > 0$, $E \in (0, D)$ *and* $F := D - E$. *If, for a choice* $\Theta_i \in \{\Theta, \Xi\}$, $1 \le i < 3$, *it is*

$$S_F := \big\| \mathbf{E}_F(\Theta_2, \Theta_3) \big\|_\infty = \sup_{\theta_2 \in \Theta_2} \sum_{\theta_3 \in \Theta_3} e^{-F \varrho_{\Theta_2 \times \Theta_3}(\theta_2,\theta_3)} < \infty, \quad (2.2.21)$$

then the matrix product

$$\mathbf{A} := \mathbf{E}_E(\Theta_1, \Theta_3) \mathbf{E}_D(\Theta_3, \Theta_2)$$

fulfills the off–diagonal decay estimate

$$|a_{\theta_1,\theta_2}| \le S_F e^{-E \varrho_{\Theta_1 \times \Theta_2}(\theta_1,\theta_2)}, \quad \text{for all } \theta_1 \in \Theta_1, \ \theta_2 \in \Theta_2. \quad (2.2.22)$$

If, moreover,

$$S_F' := \big\| \mathbf{E}_F(\Theta_1, \Theta_3) \big\|_\infty = \sup_{\theta_1 \in \Theta_1} \sum_{\theta_3 \in \Theta_3} e^{-F \varrho_{\Theta_1 \times \Theta_3}(\theta_1,\theta_3)} < \infty, \quad (2.2.23)$$

then an analogous estimate holds for reversed roles of E *and* D

$$\big| \big(\mathbf{E}_D(\Theta_1, \Theta_3) \mathbf{E}_E(\Theta_3, \Theta_2) \big)_{\theta_1,\theta_2} \big| \le S_F' e^{-E \varrho_{\Theta_1 \times \Theta_2}(\theta_1,\theta_2)}, \quad \text{for all } \theta_1 \in \Theta_1, \ \theta_2 \in \Theta_2.$$

If $\Theta_1 = \Theta_2 = \Theta_3$, *then powers* $\mathbf{E}_D(\Theta_1, \Theta_1)^n =: (a_{\theta_1,\theta_2}^{(n)})$ *have the off–diagonal decay*

$$|a_{\theta_1,\theta_2}^{(n)}| \le S_F^{n-1} e^{-E \varrho_{\Theta_1 \times \Theta_1}(\theta_1,\theta_2)}, \quad \text{for all } \theta_1, \theta_2 \in \Theta_1. \quad (2.2.24)$$

Proof. In [113, Section 5], only the special case $\Theta = \Xi$ was considered, but the proof also works in the general setting. Using (2.2.16c) and (2.2.21), a direct calculation yields (2.2.22):

$$
\begin{aligned}
|a_{\theta_1,\theta_2}| &= \sum_{\theta_3 \in \Theta_3} e^{-E\varrho_{\Theta_1 \times \Theta_3}(\theta_1,\theta_3)} e^{-D\varrho_{\Theta_2 \times \Theta_3}(\theta_2,\theta_3)} \\
&\leq \sum_{\theta_3 \in \Theta_3} e^{-E(\varrho_{\Theta_1 \times \Theta_2}(\theta_1,\theta_2) - \varrho_{\Theta_2 \times \Theta_3}(\theta_2,\theta_3))} e^{-D\varrho_{\Theta_2 \times \Theta_3}(\theta_2,\theta_3)} \\
&= e^{-E\varrho_{\Theta_1 \times \Theta_2}(\theta_1,\theta_2)} \sum_{\theta_3 \in \Theta_3} e^{-(D-E)\varrho_{\Theta_2 \times \Theta_3}(\theta_2,\theta_3)} \\
&\leq S_F e^{-E\varrho_{\Theta_1 \times \Theta_2}(\theta_1,\theta_2)}.
\end{aligned}
$$

For interchanged roles of D and E and $\mathbf{A} = \mathbf{E}_D(\Theta_1, \Theta_3)\mathbf{E}_E(\Theta_3, \Theta_2)$, one calculates analogously

$$
\begin{aligned}
|a_{\theta_1,\theta_2}| &= \sum_{\theta_3 \in \Theta_3} e^{-D\varrho_{\Theta_1 \times \Theta_3}(\theta_1,\theta_3)} e^{-E\varrho_{\Theta_2 \times \Theta_3}(\theta_2,\theta_3)} \\
&\leq \sum_{\theta_3 \in \Theta_3} e^{-D\varrho_{\Theta_1 \times \Theta_3}(\theta_1,\theta_3)} e^{-E(\varrho_{\Theta_1 \times \Theta_3}(\theta_1,\theta_3) - \varrho_{\Theta_1 \times \Theta_2}(\theta_1,\theta_2))} \\
&= e^{-E\varrho_{\Theta_1 \times \Theta_2}(\theta_1,\theta_2)} \sum_{\theta_3 \in \Theta_3} e^{-(D-E)\varrho_{\Theta_1 \times \Theta_3}(\theta_1,\theta_3)} \\
&\leq S'_F e^{-E\varrho_{\Theta_1 \times \Theta_2}(\theta_1,\theta_2)}.
\end{aligned}
$$

(2.2.24) follows by induction. □

Under the same additional condition (2.2.21), the matrix class of Lemarié–like decay for $\Theta = \Xi$ is nearly inverse–closed. This was shown in [113, Théorème 5] for the classical Lemarié metric and can be transferred to the generalized case:

Theorem 2.12. *Assume that ϱ is a generalized Lemarié metric for the pair (Θ, Θ), with $D > 0$, $E \in (0, D)$, $F := D - E$ and (2.2.21) holds. Let $\mathbf{M} = (m_{\theta,\xi})_{\theta,\xi \in \Theta}$ be an automorphism of $\ell_2(\Theta)$ with*

$$A\|\mathbf{c}\|_{\ell_2(\Theta)} \leq \|\mathbf{Mc}\|_{\ell_2(\Theta)} \leq B\|\mathbf{c}\|_{\ell_2(\Theta)} \tag{2.2.25}$$

and the off–diagonal decay estimate

$$|m_{\theta,\xi}| \leq Ce^{-D\varrho_{\Theta \times \Theta}(\theta,\xi)} \tag{2.2.26}$$

for some constants $A, B, C > 0$. Then the inverse $\mathbf{M}^{-1} =: (p_{\theta,\xi})_{\theta,\xi \in \Theta}$ has exponential off–diagonal decay as well:

$$|p_{\theta,\xi}| \leq C_1 e^{-D_1 \varrho_{\Theta \times \Theta}(\theta,\xi)} \tag{2.2.27}$$

for some $C_1 > 0$ and

$$D_1 = \min\left\{\frac{E}{2}, -\left\lfloor \frac{E}{2\log\left(\left(e^{D\varrho_0} + \frac{C}{B}\right)S_F\right)} \right\rfloor \log\left(1 - \frac{A}{B}\right)\right\}, \tag{2.2.28}$$

where $\varrho_0 \geq 0$ is the constant from (2.2.16a).

Proof. In [113, Théorème 5], only the case $\varrho_0 = 0$ was considered, but the proof for $\varrho_0 > 0$ is completely analogous. Without loss of generality, assume that $\mathbf{M}$ is positive self–adjoint, otherwise use $\mathbf{M}^{-1} = \mathbf{M}^*(\mathbf{M}\mathbf{M}^*)^{-1}$ and Lemma 2.11. By (2.2.25), the spectrum $\sigma(\mathbf{M})$ is contained in $[A, B]$, i.e., $\sigma(\mathbf{S}) \subset [0, 1 - \frac{A}{B}]$ for $\mathbf{S} := \mathbf{I} - \frac{1}{B}\mathbf{M}$. Moreover, $\|\mathbf{S}\| \leq 1 - \frac{A}{B} =: q < 1$, so that the Neumann series $\mathbf{M}^{-1} = \frac{1}{B}\sum_{n=0}^{\infty} \mathbf{S}^n$ can be used to estimate $|p_{\theta,\xi}|$ by the entries of $\mathbf{S}^n =: (s_{\theta,\xi}^{(n)})_{\theta,\xi\in\Theta}$. For large n, we use $|s_{\theta,\xi}^{(n)}| \leq \|\mathbf{S}^n\| \leq q^n$. For small n, we choose a number $E \in (0, D)$. By (2.2.26) and the definition of $\mathbf{S}$, we have

$$|s_{\theta,\xi}^{(1)}| \leq \delta_{\theta,\xi} + \frac{C}{B}e^{-D\varrho_\Theta\times\Theta(\theta,\xi)} \leq \left(e^{D\varrho_0} + \frac{C}{B}\right)e^{-D\varrho_\Theta\times\Theta(\theta,\xi)},$$

so that by Lemma 2.11, for $F := D - E$,

$$|s_{\theta,\xi}^{(n)}| \leq \left(e^{D\varrho_0} + \frac{C}{B}\right)^n \left((e^{-D\varrho_\Theta\times\Theta(\zeta,\xi)})_{\zeta,\xi\in\Theta}\right)_{\theta,\xi}^n \leq \left(e^{D\varrho_0} + \frac{C}{B}\right)^n S_F^{n-1}e^{-E\varrho_\Theta\times\Theta(\theta,\xi)}.$$

Hence for any $n_0 \in \mathbb{N}$ it follows that

$$
\begin{aligned}
|p_{\theta,\xi}| &\leq \frac{1}{B}\left(\delta_{\theta,\xi} + \left(\sum_{n=1}^{n_0}\left(e^{D\varrho_0} + \frac{C}{B}\right)^n S_F^{n-1}\right)e^{-E\varrho_\Theta\times\Theta(\theta,\xi)}\right) + \frac{1}{B}\sum_{n=n_0+1}^{\infty} q^n \\
&\leq \frac{1}{B}\left(e^{E\varrho_0} + \sum_{n=1}^{n_0}\left(e^{D\varrho_0} + \frac{C}{B}\right)^n S_F^{n-1}\right)e^{-E\varrho_\Theta\times\Theta(\theta,\xi)} + \frac{q^{n_0+1}}{B(1-q)} \\
&\leq \frac{e^{E\varrho_0}}{B}\left(1 + \sum_{n=1}^{n_0}\left(e^{D\varrho_0} + \frac{C}{B}\right)^n S_F^{n-1}\right)e^{-E\varrho_\Theta\times\Theta(\theta,\xi)} + \left(\frac{1}{A} - \frac{1}{B}\right)\left(1 - \frac{A}{B}\right)^{n_0}.
\end{aligned}
$$

Since

$$1 = e^{D\varrho_0}e^{-D\varrho_0} \leq e^{D\varrho_0}e^{-F\varrho_0} \leq e^{D\varrho_0}S_F,$$

we can estimate

$$
\begin{aligned}
|p_{\theta,\xi}| &\leq \frac{e^{(D+E)\varrho_0}}{B}\left(\sum_{n=0}^{n_0}\left(e^{D\varrho_0} + \frac{C}{B}\right)^n S_F^n\right)e^{-E\varrho_\Theta\times\Theta(\theta,\xi)} + \left(\frac{1}{A} - \frac{1}{B}\right)\left(1 - \frac{A}{B}\right)^{n_0} \\
&= \frac{e^{(D+E)\varrho_0}\left((e^{D\varrho_0} + \frac{C}{B})^{n_0+1}S_F^{n_0+1} - 1\right)}{B\left((e^{D\varrho_0} + \frac{C}{B})S_F - 1\right)}e^{-E\varrho_\Theta\times\Theta(\theta,\xi)} + \left(\frac{1}{A} - \frac{1}{B}\right)\left(1 - \frac{A}{B}\right)^{n_0} \\
&\leq C_1\left(\left(e^{D\varrho_0} + \frac{C}{B}\right)^{n_0}S_F^{n_0}e^{-E\varrho_\Theta\times\Theta(\theta,\xi)} + \left(1 - \frac{A}{B}\right)^{n_0}\right),
\end{aligned}
$$

where

$$C_1 := \max\left\{\frac{e^{(D+E)\varrho_0}\left(e^{D\varrho_0} + \frac{C}{B}\right)S_F}{B\left((e^{D\varrho_0} + \frac{C}{B})S_F - 1\right)}, \frac{1}{A} - \frac{1}{B}\right\}.$$

Now we choose

$$n_0 := \left\lfloor\frac{E}{2\log\left((e^{D\varrho_0} + \frac{C}{B})S_F\right)}\rho_{\Theta\times\Theta}(\theta, \xi)\right\rfloor \geq 0,$$

so that (2.2.28) follows by

$$
\begin{aligned}
&\left(\mathrm{e}^{D\varrho_0} + \frac{C}{B}\right)^{n_0} S_F^{n_0} \mathrm{e}^{-E\varrho_{\Theta\times\Theta}(\theta,\xi)} + \left(1 - \frac{A}{B}\right)^{n_0} \\
&= \; \mathrm{e}^{n_0 \log\left(\left(\mathrm{e}^{D\varrho_0} + \frac{C}{B}\right)S_F\right) - E\varrho_{\Theta\times\Theta}(\theta,\xi)} + \mathrm{e}^{n_0 \log\left(1 - \frac{A}{B}\right)} \\
&\leq \; \mathrm{e}^{-D_1 \varrho_{\Theta\times\Theta}(\theta,\xi)}.
\end{aligned}
$$

□

In the next subsection, we will also need the following result for frames that are localized to a Riesz basis. It is a slight generalization of results already presented in [41, 89].

Theorem 2.13. *Assume that ϱ is a generalized Lemarié metric for the pair (Θ, Ξ). Let $\mathcal{F} = \{f_\theta\}_{\theta\in\Theta}$ be a frame in H with canonical dual frame $\tilde{\mathcal{F}} = \{\tilde{f}_\theta\}_{\theta\in\Theta}$ and $\mathcal{G} = \{g_\xi\}_{\xi\in\Xi}$ be a Riesz basis for H with dual basis $\tilde{\mathcal{G}} = \{\tilde{g}_\xi\}_{\xi\in\Xi}$, such that $\mathcal{F}$ is ϱ-exponentially localized to $\tilde{\mathcal{G}}$,*

$$
\left|\langle f_\theta, \tilde{g}_\xi \rangle\right| \lesssim \mathrm{e}^{-\alpha\varrho_{\Theta,\Xi}(\theta,\xi)}, \quad \theta \in \Theta, \xi \in \Xi, \tag{2.2.29}
$$

for some $\alpha > 0$. Moreover, assume that $\left\|\mathbf{E}_\gamma(\Theta,\Xi)\right\| < \infty$ and $\left\|\mathbf{E}_\gamma(\Xi,\Xi)\right\| < \infty$ hold for all $0 < \gamma_0 < \gamma < \alpha$, where $\alpha > 2\gamma_0$. Then there exists $\delta \in (0, \frac{\alpha}{2} - \gamma_0)$, such that

$$
\left|\langle \tilde{f}_\theta, g_\xi \rangle\right| \lesssim \mathrm{e}^{-\delta\varrho_{\Theta,\Xi}(\theta,\xi)}, \quad \theta \in \Theta, \xi \in \Xi. \tag{2.2.30}
$$

Proof. Consider the isomorphism

$$
\tilde{G} : H \to \ell_2(\Theta), \quad \tilde{G}f := \left(\langle f, \tilde{g}_\xi \rangle\right)_{\xi\in\Xi},
$$

its adjoint

$$
\tilde{G}^* : \ell_2(\Theta) \to H, \quad \tilde{G}^*\mathbf{c} = \mathbf{c}^\top \tilde{\mathcal{G}}
$$

and the operator $\mathbf{T} := \tilde{G}S\tilde{G}^* : \ell_2(\Xi) \to \ell_2(\Xi)$, where

$$
S : H \to H, \quad Sf := \sum_{\theta\in\Theta} \langle f, f_\theta \rangle f_\theta
$$

is the frame operator associated with $\mathcal{F}$. $\mathbf{T}$ is an automorphism of $\ell_2(\Xi)$ with

$$
\mathbf{T}_{\xi_1,\xi_2} = \langle \mathbf{e}_{\xi_1}, \mathbf{T}\mathbf{e}_{\xi_2} \rangle_{\ell_2(\Xi)} = \sum_{\xi\in\Xi} \langle \tilde{g}_{\xi_2}, f_\xi \rangle \langle f_\xi, \tilde{g}_{\xi_1} \rangle.
$$

Moreover, since by assumption $\left\|\mathbf{E}_\gamma(\Xi,\Theta)\right\|_\infty < \infty$ for $0 < \gamma_0 < \gamma < \alpha$, Lemma 2.11 yields the componentwise estimate

$$
|\mathbf{T}| \leq \mathbf{E}_{\alpha-\gamma}(\Xi,\Theta)\mathbf{E}_\alpha(\Theta,\Xi) \leq \left\|\mathbf{E}_\gamma(\Xi,\Theta)\right\|_\infty \mathbf{E}_{\alpha-\gamma}(\Xi,\Xi).
$$

Since $\left\|\mathbf{E}_{\alpha-\gamma}(\Xi,\Xi)\right\|_\infty < \infty$ holds for $0 < \gamma_0 < \alpha - \gamma$, an application of Theorem 2.12 with $D = \alpha - \gamma$, $\gamma_0 < F < \alpha - \gamma$ and $E = D - F = \alpha - \gamma - F \in (0, \alpha - \gamma - \gamma_0)$ guarantees the existence of a $0 < \delta \leq \frac{E}{2} < \frac{\alpha-\gamma-\gamma_0}{2}$, such that $|\mathbf{T}^{-1}| \lesssim \mathbf{E}_\delta(\Xi,\Xi)$.

Since by a straightforward computation, it is

$$
\begin{aligned}
\langle f_\theta, \tilde{g}_\xi \rangle &= \langle SS^{-1} f_\theta, \tilde{g}_\xi \rangle = \Big\langle \sum_{\mu \in \Theta} \langle S^{-1} f_\theta, f_\mu \rangle f_\mu, \tilde{g}_\xi \Big\rangle \\
&= \sum_{\mu \in \Theta} \langle S^{-1} f_\theta, f_\mu \rangle \langle f_\mu, \tilde{g}_\xi \rangle = \sum_{\mu \in \Theta} \langle \tilde{f}_\theta, f_\mu \rangle \langle f_\mu, \tilde{g}_\xi \rangle \\
&= \sum_{\mu \in \Theta} \Big(\sum_{\nu \in \Xi} \langle \tilde{f}_\theta, g_\nu \rangle \langle \tilde{g}_\nu, f_\mu \rangle \Big) \langle f_\mu, \tilde{g}_\xi \rangle \\
&= (\mathbf{AT})_{\theta,\xi},
\end{aligned}
$$

where

$$
\mathbf{A} := \big(\langle \tilde{f}_\theta, g_\xi \rangle \big)_{\theta \in \Theta, \xi \in \Xi} = \Big(\sum_{\mu \in \Theta} \langle f_\theta, \tilde{g}_\mu \rangle (\mathbf{T}^{-1})_{\mu,\xi} \Big)_{\theta \in \Theta, \xi \in \Xi},
$$

another application of Lemma 2.11 yields

$$
|\mathbf{A}| \lesssim \mathbf{E}_\alpha(\Theta, \Xi) |\mathbf{T}^{-1}| \lesssim \mathbf{E}_\alpha(\Theta, \Xi) \mathbf{E}_\delta(\Xi, \Xi) \leq \|\mathbf{E}_{\alpha-\delta}(\Theta, \Xi)\|_\infty \mathbf{E}_\delta(\Theta, \Xi).
$$

Here we have used that

$$
\alpha - \delta > \alpha - \frac{\alpha - \gamma - \gamma_0}{2} = \frac{\alpha + \gamma + \gamma_0}{2} > \gamma_0,
$$

so that $\|\mathbf{E}_{\alpha-\delta}(\Theta, \Xi)\|_\infty < \infty$ by assumption. $\qquad\square$

Remark 2.14. *In order to obtain a strong localization of $\tilde{\mathcal{F}}$ against the Riesz basis $\mathcal{G}$, it is obvious that one needs $\alpha \gg 2\gamma_0$ in Theorem 2.13.*

2.2.3 Proof of the Gelfand Frame Property

As already discussed in Subsection 2.2.1, our aim is to show that the aggregated system $\Psi = \{\psi_\lambda\}_{\lambda \in \mathcal{J}}$ from (2.2.8) is not only a frame in $L_2(\Omega)$, but also a Gelfand frame for some Gelfand triple $(H_0^s(\Omega), L_2(\Omega), H^{-s}(\Omega))$. In this subsection, we will see how the concept of localized frames may come into play as one possible tool to do so.

First of all, we shall define an appropriate Lemarié metric for the pair of index sets $(\mathcal{J}, \mathcal{J})$, mimicking the distance function (2.2.13) to some extent. For parameters $\beta, \sigma > 0$ and all $\lambda = (i, j, \mathbf{e}, \mathbf{k}), \lambda' = (i', j', \mathbf{e}', \mathbf{k}') \in \mathcal{J}$, let us define the function

$$
\begin{aligned}
\varrho_1(\lambda, \lambda') := &\frac{\beta}{\sigma} \log \Big(1 + 2^{\min\{j,j'\}} \big\| \kappa_i(2^{-j}\mathbf{k}) - \kappa_{i'}(2^{-j'}\mathbf{k}') \big\| \Big) \\
&+ |j - j'| \log 2 + \frac{9\beta}{2\sigma} \log 2,
\end{aligned} \tag{2.2.31}
$$

i.e., for $\alpha > 0$ we have

$$
\mathrm{e}^{-\alpha \varrho_1(\lambda, \lambda')} = 2^{-9\alpha\beta/(2\sigma)} \Big(1 + 2^{\min\{j,j'\}} \big\| \kappa_i(2^{-j}\mathbf{k}) - \kappa_{i'}(2^{-j'}\mathbf{k}') \big\| \Big)^{-\alpha\beta/\sigma} 2^{-\alpha|j-j'|}. \tag{2.2.32}
$$

Note that the wavelet type parameters $\mathbf{e}, \mathbf{e}'$ do not enter the definition of $\varrho_1(\lambda, \lambda')$. Essentially, this is due to the fact that the supports of ψ_λ and $\psi_{\lambda'}$ only weakly depend on the wavelet type.

For the further analysis, we will need the following lemma, see [79]:

Lemma 2.15. *On the real upper half plane* $\mathbb{H}^d := \mathbb{R}^d \times \mathbb{R}_+$, *define the function*

$$\varrho_P(\mathbf{y}, \mathbf{y}') := \text{Artanh}\,\vartheta = \frac{1}{2}\log\frac{1+\vartheta}{1-\vartheta}, \quad \mathbf{y}, \mathbf{y}' \in \mathbb{H}^d \qquad (2.2.33)$$

where

$$\vartheta := \vartheta(\mathbf{y}, \mathbf{y}') := \frac{\|\mathbf{y} - \mathbf{y}'\|}{\|\mathbf{y} - \overline{\mathbf{y}}'\|} \in [0, 1) \qquad (2.2.34)$$

and $\overline{\mathbf{y}} := (\mathbf{x}, -t)$ *for* $\mathbf{y} = (\mathbf{x}, t) \in \mathbb{H}$. *Then* $(\mathbb{H}^d, \varrho_P)$ *is a metric space.*

Remark 2.16. *The metric* ϱ_P *on* $\mathbb{H}^d$ *is a straightforward generalization of the* Poincaré *metric* δ_P *on the complex upper half plane* $\mathbb{H} := \{z \in \mathbb{C} : \text{Im}\, z > 0\}$

$$\delta_P(z_1, z_2) := \text{Artanh}\left|\frac{z_2 - z_1}{z_2 - \overline{z_1}}\right|, \quad z_1, z_2 \in \mathbb{H} \qquad (2.2.35)$$

to the real upper half plane $\mathbb{H}^d$. *The distance function* δ_P *is indeed a metric on the* Poincaré *half plane* $\mathbb{H}$, *since it is the composition* $\delta_P(z_1, z_2) = \delta_h(\varphi(z_1), \varphi(z_2))$ *of the hyperbolic metric* δ_h *in the open unit disc* $\mathbb{D} := \{z \in \mathbb{C} : |z| < 1\}$

$$\delta_h(z_1, z_2) = \inf_{\substack{\gamma:[0,1]\to\mathbb{D}, \\ \gamma(0)=z_1, \gamma(1)=z_2}} \int_0^1 \frac{|\gamma'(t)|}{1 - |\gamma(t)|^2}\, dt = \text{Artanh}\left|\frac{z_2 - z_1}{1 - \overline{z_1}z_2}\right|, \quad z_1, z_2 \in \mathbb{D}$$

and the Möbius *transformation* $\varphi : \mathbb{H} \to \mathbb{D}$, $\varphi(z) = \frac{iz+1}{z+i}$, *see* [76, 111] *for details.*

Using Lemma 2.15, the generalized Lemarié metric axioms can be checked easily:

Lemma 2.17. *For* $\sigma > 0$ *and* $0 < \beta < 2\sigma$, *the tuple* $\varrho = (\varrho_1)$ *is a generalized* Lemarié *metric for the tuple of index sets* $(\mathcal{J}, \mathcal{J})$.

Proof. Property (2.2.16a) follows from $\varrho_1(\lambda, \lambda) = \frac{9\beta}{2\sigma}\log 2$, (2.2.16b) is trivial. For the triangle inequality (2.2.16c), we will use an analogous argument as in [79], which is based on the metric (2.2.33) on the upper half plane $\mathbb{H}^{d+1}$. Like in [79], one observes that

$$\left(\frac{1+\vartheta}{1-\vartheta}\right)^{1/2} = \frac{1+\vartheta}{2}\left(\frac{|t'+t|^2}{t't}\right)^{1/2}\left(1 + \frac{\|\mathbf{x}' - \mathbf{x}\|^2}{|t'+t|^2}\right)^{1/2}. \qquad (2.2.36)$$

We have the equivalence

$$\frac{1}{2}\left(\frac{1+\vartheta}{1-\vartheta}\right)^{1/2} \leq \max\left\{\sqrt{\frac{t'}{t}}, \sqrt{\frac{t}{t'}}\right\}\left(1 + \frac{\|\mathbf{x}' - \mathbf{x}\|}{\max\{t, t'\}}\right) \leq \sqrt{32}\left(\frac{1+\vartheta}{1-\vartheta}\right)^{1/2},$$
$$\qquad (2.2.37)$$

since by (2.2.34) and (2.2.36)

$$
\begin{aligned}
\frac{1+\vartheta}{1-\vartheta} &= \frac{(1+\vartheta)^2}{4}\left(\frac{|t'+t|^2}{t't}\right)\left(1+\frac{\|\mathbf{x}'-\mathbf{x}\|^2}{|t'+t|^2}\right) \\
&\leq \left(\frac{t'}{t}+2+\frac{t}{t'}\right)\left(1+\frac{\|\mathbf{x}'-\mathbf{x}\|}{|t'+t|}\right)^2 \\
&\leq 4\max\left\{\frac{t'}{t},\frac{t}{t'}\right\}\left(1+\frac{\|\mathbf{x}'-\mathbf{x}\|}{\max\{t,t'\}}\right)^2 \\
&= 4\left(\max\left\{\sqrt{\frac{t'}{t}},\sqrt{\frac{t}{t'}}\right\}\right)^2\left(1+\frac{\|\mathbf{x}'-\mathbf{x}\|}{\max\{t,t'\}}\right)^2 \\
&\leq 16(1+\vartheta)^2\left(\sqrt{\frac{t'}{t}}+\sqrt{\frac{t}{t'}}\right)^2\left(1+\frac{\|\mathbf{x}'-\mathbf{x}\|}{|t'+t|}\right)^2 \\
&\leq 32(1+\vartheta)^2\left(\frac{t'}{t}+2+\frac{t}{t'}\right)\left(1+\frac{\|\mathbf{x}'-\mathbf{x}\|^2}{|t'+t|^2}\right) \\
&= 128\frac{1+\vartheta}{1-\vartheta}.
\end{aligned}
$$

In the following, we use (2.2.37) at $(\mathbf{x},t)=\big(\kappa_i(2^{-j}\mathbf{k}),2^{-j}\big)$. Since (2.2.31) yields

$$
\begin{aligned}
&\varrho_1\big((i,j,\mathbf{k}),(i',j',\mathbf{k}')\big) \\
&= \frac{\beta}{\sigma}\log\left(1+2^{\min\{j,j'\}}\big\|\kappa_i(2^{-j}\mathbf{k})-\kappa_{i'}(2^{-j'}\mathbf{k}')\big\|\right)+|j-j'|\log 2+\frac{9\beta}{2\sigma}\log 2 \\
&= \frac{\beta}{\sigma}\log\left(2^{|j-j'|(\sigma/\beta-1/2)}2^{|j-j'|/2}\left(1+\frac{\big\|\kappa_i(2^{-j}\mathbf{k})-\kappa_{i'}(2^{-j'}\mathbf{k}')\big\|}{\max\{2^{-j},2^{-j'}\}}\right)\right)+\frac{9\beta}{2\sigma}\log 2,
\end{aligned}
$$

(2.2.37) and the metric properties of ϱ_{P} imply for $\beta<2\sigma$

$$
\begin{aligned}
&\varrho_1\big((i,j,\mathbf{k}),(i'',j'',\mathbf{k}'')\big) \\
&= \frac{\beta}{\sigma}\log\left(2^{|j-j''|(\sigma/\beta-1/2)}2^{|j-j''|/2}\left(1+\frac{\big\|\kappa_i(2^{-j}\mathbf{k})-\kappa_{i''}(2^{-j''}\mathbf{k}'')\big\|}{\max\{2^{-j},2^{-j''}\}}\right)\right)+\frac{9\beta}{2\sigma}\log 2 \\
&\leq \frac{\beta}{\sigma}\log\left(\sqrt{32}\left(\frac{1+\vartheta\big((\mathbf{x},t),(\mathbf{x}'',t'')\big)}{1-\vartheta\big((\mathbf{x},t),(\mathbf{x}'',t'')\big)}\right)^{1/2}\right)+|j-j''|\left(1-\frac{\beta}{2\sigma}\right)\log 2+\frac{9\beta}{2\sigma}\log 2 \\
&\leq \frac{\beta}{\sigma}\varrho_{\mathrm{P}}\big((\kappa_i(2^{-j}\mathbf{k}),2^{-j}),(\kappa_{i'}(2^{-j'}\mathbf{k}'),2^{-j'})\big)+|j-j'|\left(1-\frac{\beta}{2\sigma}\right)\log 2 \\
&\quad +\frac{\beta}{\sigma}\varrho_{\mathrm{P}}\big((\kappa_{i'}(2^{-j'}\mathbf{k}'),2^{-j'}),(\kappa_{i''}(2^{-j''}\mathbf{k}''),2^{-j''})\big)+|j'-j''|\left(1-\frac{\beta}{2\sigma}\right)\log 2+\frac{7\beta}{\sigma}\log 2 \\
&\leq \frac{\beta}{\sigma}\log\left(2^{|j-j'|/2}\left(1+\frac{\big\|\kappa_i(2^{-j}\mathbf{k})-\kappa_{i'}(2^{-j'}\mathbf{k}')\big\|}{\max\{2^{-j},2^{-j'}\}}\right)\right)+|j-j'|\left(1-\frac{\beta}{2\sigma}\right)\log 2 \\
&\quad +\frac{\beta}{\sigma}\log\left(2^{|j'-j''|/2}\left(1+\frac{\big\|\kappa_{i'}(2^{-j'}\mathbf{k}')-\kappa_{i''}(2^{-j''}\mathbf{k}'')\big\|}{\max\{2^{-j'},2^{-j''}\}}\right)\right)+|j'-j''|\left(1-\frac{\beta}{2\sigma}\right)\log 2 \\
&\quad +\frac{9\beta}{\sigma}\log 2 \\
&= \varrho_1\big((i,j,\mathbf{k}),(i',j',\mathbf{k}')\big)+\varrho_1\big((i',j',\mathbf{k}'),(i'',j'',\mathbf{k}'')\big).
\end{aligned}
$$

$\square$

Since, in the following, we will frequently use Lemma 2.11 and Theorem 2.12, we have to clarify first for which range of $\gamma > 0$ the condition (2.2.21) holds. Using a Riemann–type argument, observe that for $\lambda \in \mathcal{J}$, $j' \geq j_0$ and any $r > d$, we can assume that the estimates

$$\sum_{|\lambda'|=j'} \left(1 + 2^{\min\{j,j'\}} \left\| \kappa_i(2^{-j}\mathbf{k}) - \kappa_{i'}(2^{-j'}\mathbf{k}') \right\| \right)^{-r} \lesssim 2^{-j'} \int_{\mathbb{R}^d} \left(1 + 2^{\min\{j,j'\}} \|\mathbf{x}\| \right)^{-r} d\mathbf{x}$$

and therefore, by estimating the integral expression further, also

$$\sum_{|\lambda'|=j'} \left(1 + 2^{\min\{j,j'\}} \left\| \kappa_i(2^{-j}\mathbf{k}) - \kappa_{i'}(2^{-j'}\mathbf{k}') \right\| \right)^{-r} \lesssim 2^{d \max\{0, j'-j\}} \tag{2.2.38}$$

hold, where the constants involved only depend on r. As an immediate consequence, we get the following lemma:

Lemma 2.18. *For the generalized Lemarié metric* $\varrho = (\varrho_1)$ *with parameters* $\sigma > 0$ *and* $0 < \beta < 2\sigma$, *it is* $\|\mathbf{E}_\gamma\|_\infty < \infty$ *for all* $\gamma > \max\{\frac{\sigma d}{\beta}, d\}$.

Proof. Note that for $0 < \beta < 2\sigma$, ϱ is indeed a Lemarié metric by Lemma 2.17. Since $\frac{\beta\gamma}{\sigma} > d$, (2.2.38) yields

$$\sum_{|\lambda'|=j'} \left(1 + 2^{\min\{j,j'\}} \left\| \kappa_i(2^{-j}\mathbf{k}) - \kappa_{i'}(2^{-j'}\mathbf{k}') \right\| \right)^{-\beta\gamma/\sigma} \lesssim 2^{d \max\{0, j'-j\}}$$

and therefore

$$\sum_{\lambda' \in \mathcal{J}} e^{-\gamma\varrho_1(\lambda, \lambda')}$$

$$= 2^{-9\beta\gamma/(2\sigma)} \sum_{j' \geq j_0} 2^{-\gamma|j'-j|} \sum_{|\lambda'|=j'} \left(1 + 2^{\min\{j,j'\}} \left\| \kappa_i(2^{-j}\mathbf{k}) - \kappa_{i'}(2^{-j'}\mathbf{k}') \right\| \right)^{-\beta\gamma/\sigma}$$

$$\lesssim \sum_{j' \geq j_0} 2^{-\gamma|j'-j|+d\max\{0, j'-j\}}$$

$$\leq \sum_{j' \geq j_0} 2^{-(\gamma-d)|j'-j|}.$$

Since $\gamma > d$ by assumption, the latter sum can be estimated uniformly in λ by

$$\sum_{j' \geq j_0} 2^{-(\gamma-d)|j'-j|} = \sum_{\substack{j' \geq j_0 \\ j' \leq j}} 2^{-(\gamma-d)(j-j')} + \sum_{\substack{j' \geq j_0 \\ j' > j}} 2^{-(\gamma-d)(j'-j)}$$

$$\leq 2 \sum_{j'' \geq 0} 2^{-(\gamma-d)j''}$$

$$= \frac{2}{1 - 2^{d-\gamma}} =: C_\gamma < \infty,$$

so that $\|\mathbf{E}_\gamma\|_\infty \leq C_\gamma < \infty$. $\qquad\square$

Remark 2.19. *(i) The result of Lemma 2.18 resembles much that of [89, Lemma 2.1]. There, the special case $\varrho_1(\lambda, \lambda') = \log\left(1 + \|\lambda - \lambda'\|\right)$ for a relatively separated subset $\mathcal{J} \subset \mathbb{R}^d$ was considered, with $\|\mathbf{E}_\gamma\|_\infty < \infty$ for $\gamma > d$. In fact, it is possible to exchange the integral argument (2.2.38) in Lemma 2.18 by the techniques from [89], resulting in an estimate of the form*

$$\sum_{|\lambda'|=j'} \left(1 + 2^{\min\{j,j'\}}\big\|\kappa_i(2^{-j}\mathbf{k}) - \kappa_{i'}(2^{-j'}\mathbf{k}')\big\|\right)^{-r} \lesssim 2^{r\max\{0,j'-j\}} \qquad (2.2.39)$$

for all $r > d$, which is weaker than (2.2.38) but still sufficient to prove the boundedness of $\|\mathbf{E}_\gamma\|_\infty$ in the range $\gamma > \max\{\frac{\sigma d}{\beta}, d\}$. However, we shall need the stronger estimate (2.2.38) later on.

(ii) Since in the Lemarié metric case, the range for boundedness of $\|\mathbf{E}_\gamma\|_\infty$ depends on the choice of β and σ and not only on the global constant d, we will have to keep track of this dependence when it comes to any application of Lemma 2.11.

For the further discussion, note that any sequence $\mathbf{c} \in \ell_p(\mathcal{J})$ can be uniquely resorted as an M–tuple $\mathbf{c} = (\mathbf{c}^{(1)}, \ldots, \mathbf{c}^{(M)}) \in \ell_p(\mathcal{J}^\square)^M$ with $c_{(i,\mu)} = c_\mu^{(i)}$ and equivalent norms

$$\|\mathbf{c}\|_{\ell_p(\mathcal{J})} \approx \|(\mathbf{c}^{(1)}, \ldots, \mathbf{c}^{(M)})\|_{\ell_p(\mathcal{J}^\square)^M} := \sum_{i=1}^M \|\mathbf{c}^{(i)}\|_{\ell_p(\mathcal{J}^\square)}.$$

Since the entries of the diagonal matrix $\mathbf{D} = (\delta_{\lambda,\lambda'} 2^{|\lambda|})_{\lambda,\lambda' \in \mathcal{J}}$ do not depend on the patch numbers, this resorting can also be applied in the situation of weighted sequence spaces. So, with a slight abuse of notation concerning a double use of $\mathbf{D}$ as a matrix both over $\mathcal{J}$ and $\mathcal{J}^\square$, we also have the equivalence

$$\|\mathbf{c}\|_{\ell_{p,\mathbf{D}^s}(\mathcal{J})} \approx \|(\mathbf{c}^{(1)}, \ldots, \mathbf{c}^{(M)})\|_{\ell_{p,\mathbf{D}^s}(\mathcal{J}^\square)^M} := \sum_{i=1}^M \|\mathbf{c}^{(i)}\|_{\ell_{p,\mathbf{D}^s}(\mathcal{J}^\square)}, \qquad (2.2.40)$$

where the constants involved do not depend on $s \geq 0$.

With this notation, matrices with ϱ–exponential decay yield bounded operators on weighted ℓ_p spaces:

Lemma 2.20. *For parameters $\beta, \sigma > 0$, let $\varrho_1 : \mathcal{J} \times \mathcal{J} \to \mathbb{R}_+^0$ be given by (2.2.31). Then, for $\gamma > \max\{\frac{\sigma d}{\beta}, d\}$, the matrix $\mathbf{E}_\gamma := \mathbf{E}_\gamma(\mathcal{J}, \mathcal{J}) = (\mathrm{e}^{-\gamma \varrho_1(\lambda, \lambda')})_{\lambda,\lambda' \in \mathcal{J}}$ is a bounded operator from $\ell_{p,\mathbf{D}^s}(\mathcal{J})$ to $\ell_{p,\mathbf{D}^s}(\mathcal{J})$ for all $p \in [1, \infty]$ and for any $s \in (0, \gamma - d)$.*

Proof. Similar to the proof of the Schur lemma, we prove the boundedness of $\mathbf{E}_\gamma$ on $\ell_{1,\mathbf{D}^s}(\mathcal{J})$ and on $\ell_{\infty,\mathbf{D}^s}(\mathcal{J})$, and then conclude by interpolation of weighted sequence spaces, see [11] for details. For the boundedness on $\ell_{1,\mathbf{D}^s}(\mathcal{J})$, we compute

$$\|\mathbf{E}_\gamma \mathbf{c}\|_{\ell_{1,\mathbf{D}^s}(\mathcal{J})}$$

$$\leq \sum_{\lambda,\lambda' \in \mathcal{J}} e^{-\gamma \varrho_1(\lambda,\lambda')} |c_{\lambda'}| 2^{sj}$$

$$= \sum_{\lambda,\lambda' \in \mathcal{J}} 2^{-\gamma|j-j'|} 2^{sj} \left(1 + 2^{\min\{j,j'\}} \left\| \kappa_i(2^{-j}\mathbf{k}) - \kappa_{i'}(2^{-j'}\mathbf{k}') \right\| \right)^{-\beta\gamma/\sigma} |c_{\lambda'}|$$

$$= \sum_{j,j' \geq j_0} 2^{-\gamma|j-j'|} 2^{sj} \sum_{|\lambda'|=j'} |c_{\lambda'}| \sum_{|\lambda|=j} \left(1 + 2^{\min\{j,j'\}} \left\| \kappa_i(2^{-j}\mathbf{k}) - \kappa_{i'}(2^{-j'}\mathbf{k}') \right\| \right)^{-\beta\gamma/\sigma}.$$

Since for fixed $\lambda' \in \mathcal{J}$ and $j \geq j_0$, the estimate (2.2.38) yields

$$\sum_{|\lambda|=j} \left(1 + 2^{\min\{j,j'\}} \left\| \kappa_i(2^{-j}\mathbf{k}) - \kappa_{i'}(2^{-j'}\mathbf{k}') \right\| \right)^{-\beta\gamma/\sigma} \lesssim 2^{d\max\{0,j-j'\}},$$

with a constant independent from λ' and j, we get

$$\|\mathbf{E}_\gamma \mathbf{c}\|_{\ell_{1,\mathbf{D}^s}(\mathcal{J})} \lesssim \sum_{j,j' \geq j_0} 2^{-\gamma|j-j'|} 2^{sj} 2^{d\max\{0,j-j'\}} \sum_{|\lambda'|=j'} |c_{\lambda'}|$$

$$\leq \sum_{j,j' \geq j_0} 2^{-(\gamma-d)|j-j'|} 2^{sj} \sum_{|\lambda'|=j'} |c_{\lambda'}|$$

$$\leq \sum_{j' \geq j_0} 2^{sj'} \sum_{|\lambda'|=j'} |c_{\lambda'}| \sum_{j \geq j_0} 2^{-(\gamma-d-s)|j-j'|}$$

$$\lesssim \|\mathbf{c}\|_{\ell_{1,\mathbf{D}^s}(\mathcal{J})},$$

where we have used the assumption that $s < \gamma - d$. For the boundedness of $\mathbf{E}_\gamma$ on $\ell_{\infty,\mathbf{D}^s}(\mathcal{J})$, observe that

$$\|\mathbf{E}_\gamma \mathbf{c}\|_{\ell_{\infty,\mathbf{D}^s}(\mathcal{J})}$$

$$\leq \sup_{\lambda \in \mathcal{J}} \sum_{\lambda' \in \mathcal{J}} e^{-\gamma \varrho_1(\lambda,\lambda')} |c_{\lambda'}| 2^{sj}$$

$$= \sup_{j \geq j_0} \sum_{j' \geq j_0} 2^{-\gamma|j-j'|} 2^{sj} \sup_{|\lambda|=j} \sum_{|\lambda'|=j'} \left(1 + 2^{\min\{j,j'\}} \left\| \kappa_i(2^{-j}\mathbf{k}) - \kappa_{i'}(2^{-j'}\mathbf{k}') \right\| \right)^{-\beta\gamma/\sigma} |c_{\lambda'}|.$$

Using again the estimate (2.2.38) yields

$$\sup_{|\lambda|=j} \sum_{|\lambda'|=j'} \left(1 + 2^{\min\{j,j'\}} \left\| \kappa_i(2^{-j}\mathbf{k}) - \kappa_{i'}(2^{-j'}\mathbf{k}') \right\| \right)^{-\beta\gamma/\sigma} |c_{\lambda'}| \lesssim 2^{d\max\{0,j'-j\}} \sup_{|\lambda'|=j'} |c_{\lambda'}|$$

and hence

$$\|\mathbf{E}_\gamma \mathbf{c}\|_{\ell_{\infty,\mathbf{D}^s}(\mathcal{J})} \lesssim \sup_{j \geq j_0} \sum_{j' \geq j_0} 2^{-\gamma|j-j'|} 2^{sj} 2^{d\max\{0,j'-j\}} \sup_{|\lambda'|=j'} |c_{\lambda'}|$$

$$\leq \sup_{j \geq j_0} \sum_{j' \geq j_0} 2^{-(\gamma-d-s)|j-j'|} \|\mathbf{c}\|_{\ell_{\infty,\mathbf{D}^s}(\mathcal{J})}$$

$$\lesssim \|\mathbf{c}\|_{\ell_{\infty,\mathbf{D}^s}(\mathcal{J})},$$

which proves the claim. Note that here, in contrast to Lemma 2.18, we did not have to use any further restriction on the Lemarié metric parameter $\beta > 0$. $\square$

Theorem 2.21. *Let $\varrho = (\varrho_l)$ be the generalized Lemarié metric from (2.2.31) for the parameters $\sigma > 0$ and $0 < \beta < 2\sigma$. Moreover, assume that $\alpha > \max\{\frac{\sigma d}{\beta}, d\}$ and, for $s \in (0, \alpha - d)$, assume that $\Psi^{\square}$ is a Gelfand frame in $(H_0^s(\square), L_2(\square), H^{-s}(\square))$ for the Gelfand triple of sequence spaces $(\ell_{2,\mathbf{D}^s}(\mathcal{J}^{\square}), \ell_2(\mathcal{J}^{\square}), \ell_{2,\mathbf{D}^{-s}}(\mathcal{J}^{\square}))$. Furthermore, assume that Ψ is localized against its global canonical dual $\widetilde{\Psi}$*

$$\left|\langle \psi_\lambda, \widetilde{\psi_{\lambda'}} \rangle\right| \lesssim \mathrm{e}^{-\alpha \varrho_l(\lambda, \lambda')}, \quad \lambda, \lambda' \in \mathcal{J}. \tag{2.2.41}$$

If, additionally, the overlapping decomposition (2.2.1) of Ω satisfies the estimate

$$\|u\|_{H^s(\Omega)} \approx \inf_{\substack{u_i \in H_0^s(\Omega_i) \\ u = \sum_{i=1}^M E_i u_i}} \sum_{i=1}^M \|u_i\|_{H^s(\Omega_i)} \tag{2.2.42}$$

uniformly in $u \in H_0^s(\Omega)$, then Ψ is a Gelfand frame for $\left(H_0^s(\Omega), L_2(\Omega), H^{-s}(\Omega)\right)$ with respect to the Gelfand triple of sequence spaces $(\ell_{2,\mathbf{D}^s}(\mathcal{J}), \ell_2(\mathcal{J}), \ell_{2,\mathbf{D}^{-s}}(\mathcal{J}))$.

Proof. Ψ is a Hilbert frame by Lemma 2.7. From Lemma 2.5, we know that the local systems $\Psi^{(i)}$ are Hilbert frames in $L_2(\Omega_i)$ with canonical dual $\widetilde{\Psi}^{(i)}$, and they are Gelfand frames for $(H_0^s(\Omega_i), L_2(\Omega_i), H^{-s}(\Omega_i))$ for the Gelfand triple of sequence spaces $(\ell_{2,\mathbf{D}^s}(\mathcal{J}^{\square}), \ell_2(\mathcal{J}^{\square}), \ell_{2,\mathbf{D}^{-s}}(\mathcal{J}^{\square}))$. Hence, the corresponding Gelfand frame operators

$$\tilde{F}^{(i)} : H_0^s(\Omega_i) \to \ell_{2,\mathbf{D}^s}(\mathcal{J}^{\square}), \quad g \mapsto \langle g, \tilde{\Psi}^{(i)} \rangle_{H_0^s(\Omega_i) \times H^{-s}(\Omega_i)}^\top \tag{2.2.43}$$

and

$$(F^{(i)})^* : \ell_{2,\mathbf{D}^s}(\mathcal{J}^{\square}) \to H_0^s(\Omega_i), \quad \mathbf{c} \mapsto \mathbf{c}^\top \Psi^{(i)} \tag{2.2.44}$$

are bounded. As an immediate consequence, without using further assumptions, one can show the boundedness of F^*. In fact, for any sequence $\mathbf{c} \in \ell_{2,\mathbf{D}^s}(\mathcal{J})$, the representation

$$F^* \mathbf{c} = \sum_{i=1}^M \sum_{\mu \in \mathcal{J}^{\square}} c_\mu^{(i)} \psi_{(i,\mu)} = \sum_{i=1}^M E_i (F^{(i)})^* \mathbf{c}^{(i)},$$

the continuity of E_i and $(F^{(i)})^*$, and (2.2.40) imply

$$\|F^* \mathbf{c}\|_{H^s(\Omega)} \leq \sum_{i=1}^M \left\|(F^{(i)})^* \mathbf{c}^{(i)}\right\|_{H^s(\Omega_i)} \lesssim \sum_{i=1}^M \|\mathbf{c}^{(i)}\|_{\ell_{2,\mathbf{D}^s}(\mathcal{J}^{\square})} \approx \|\mathbf{c}\|_{\ell_{2,\mathbf{D}^s}(\mathcal{J})}.$$

Next we prove the boundedness of $\widetilde{F} : H_0^s(\Omega_i) \to \ell_{2,\mathbf{D}^s}(\mathcal{J})$, for $1 \leq i \leq M$. Given some $u \in H_0^s(\Omega)$, by assumption (2.2.42), we can assume the existence of functions $u_i \in H_0^s(\Omega_i)$ with $u = \sum_{i=1}^M E_i u_i$, such that the estimate $\|u\|_{H^s(\Omega)} \gtrsim \|u_i\|_{H^s(\Omega_i)}$ holds uniformly in u. Hence (2.2.43) yields

$$\|u\|_{H^s(\Omega)} \gtrsim \sum_{i=1}^M \|u_i\|_{H^s(\Omega_i)} \gtrsim \sum_{i=1}^M \|\tilde{F}^{(i)} u_i\|_{\ell_{2,\mathbf{D}^s}(\mathcal{J}^{\square})}. \tag{2.2.45}$$

Using the expansion of the global dual frame elements in the local dual bases

$$\widetilde{\psi}_\lambda|_{\Omega_i} = E_i^*\widetilde{\psi}_\lambda = \sum_{\mu\in\mathcal{J}^\square} \langle E_i^*\widetilde{\psi}_\lambda, \psi_{i,\mu}\rangle_{L_2(\Omega_i)} \tilde{\psi}_{i,\mu},$$

one computes the identity

$$\langle u, \widetilde{\psi}_\lambda\rangle = \sum_{i=1}^M \langle u_i, E_i^*\widetilde{\psi}_\lambda\rangle_{L_2(\Omega_i)} = \sum_{i=1}^M \sum_{\mu\in\mathcal{J}^\square} \langle u_i, \tilde{\psi}_{i,\mu}\rangle_{L_2(\Omega_i)} \langle \psi_{(i,\mu)}, \widetilde{\psi}_\lambda\rangle.$$

The matrix $\mathbf{G} = (\langle \widetilde{\psi}_\lambda, \psi_{(i,\mu)}\rangle)_{\lambda,(i,\mu)\in\mathcal{J}}$ fulfills $|\mathbf{G}| \lesssim \mathbf{E}_\alpha(\mathcal{J},\mathcal{J})$, so that $\mathbf{G}$ is bounded on $\ell_{2,\mathbf{D}^s}(\mathcal{J})$ by Lemma 2.20. Hence we have

$$\left\| \left(\langle u, \widetilde{\psi}_\lambda\rangle\right)_{\lambda\in\mathcal{J}} \right\|_{\ell_{2,\mathbf{D}^s}(\mathcal{J})} \lesssim \left\| \left(\langle u_i, \tilde{\psi}_{i,\mu}\rangle_{L_2(\Omega_i)}\right)_{(i,\mu)\in\mathcal{J}} \right\|_{\ell_{2,\mathbf{D}^s}(\mathcal{J})} \approx \sum_{i=1}^M \|\tilde{F}^{(i)}u_i\|_{\ell_{2,\mathbf{D}^s}(\mathcal{J}^\square)},$$

so that (2.2.45) yields the claim. $\square$

Remark 2.22.   *(i) In* [51], *a smooth partition of unity* $\Sigma := \{\sigma_i\}_{1\leq i\leq M}$ *subordinate to the overlapping decomposition* (2.2.1) *of* Ω *was used to prove the Gelfand frame property of* Ψ. *Unfortunately, for many interesting domains* Ω, *like the L–shaped domain, such a partition of unity does not exist. However, given such a partition of unity* $\Sigma := \{\sigma_i\}_{1\leq i\leq M}$ *with* $\sigma_i u \in H_0^s(\Omega_i)$ *and* $\|\sigma_i u\|_{H^s(\Omega_i)} \lesssim \|u\|_{H^s(\Omega)}$ *for* $s \in (0, \alpha-d)$, *it is trivial to see that* (2.2.42) *indeed holds.*

(ii) For the L–shaped domain

$$\Omega = (-1,1)^2 \setminus [0,1)^2 = (-1,0)\times(-1,1) \cup (-1,1)\times(-1,0)$$

and $s = 1$, *the condition* (2.2.42) *has been verified in* [52]. *Namely, for a smooth function* $\phi : [0, \frac{3\pi}{2}] \to \mathbb{R}_+^0$ *with* $\phi(\theta) = 1$ *for* $\theta \leq \frac{\pi}{2}$ *and* $\phi(\theta) = 0$ *for* $\theta \geq \pi$, *one can split* $H_0^1(\Omega) \ni u = u_1 + u_2$ *with* $u_1(x,y) = u(x,y)\phi(\theta(x,y))$, *where* $(r(x,y), \theta(x,y))$ *are the polar coordinates of* $(x,y) \in \Omega$ *with respect to the reentrant corner. It is* $u_i \in H_0^1(\Omega_i)$ *and one can easily show that* $\|u_i\|_{H^1(\Omega_i)} \lesssim \|u\|_{H^1(\Omega)}$, *utilizing the Lipschitz domain property of* Ω.

(iii) The condition (2.2.42) *corresponds to a so–called* stable space splitting *of* $H_0^1(\Omega)$, *see* [122, 123]. *Such decompositions of a given Hilbert space play a crucial role in domain decomposition techniques.*

It remains to show how the exponential localization (2.2.41) of Ψ against its global canonical dual $\widetilde{\Psi}$ can be realized in practice. For this purpose, we will use another localization argument. Assume that $\Psi^{\square,\circ} := \{\psi_\mu^{\square,\circ}\}_{\mu\in\mathcal{J}^{\square,\circ}}$ is a template wavelet *basis* on the unit cube, with index set $\mathcal{J}^{\square,\circ}$. We may choose $\Psi^\square = \Psi^{\square,\circ}$ both to be a wavelet basis, but since we want to leave open the possibility to choose a genuine wavelet frame $\Psi^\square$ from the very start of the construction, or to work with two different wavelet bases, let us distinguish between the two systems in the following.

Given the overlapping covering $\mathcal{C} = \{\Omega_i\}_{1 \leq i \leq M}$ of Ω, we assume that we can construct a non–overlapping auxiliary covering $\mathcal{C}^\circ = \{\Omega_i^\circ\}_{1 \leq i \leq M'}$ with diffeomorphisms $\kappa_i^\circ : \square \to \Omega_i^\circ$. Then we can define an associated aggregated system $\Psi^\circ :=\{E_i^\circ \psi_{i,\mu}^\circ\}_{(i,\mu) \in \mathcal{J}^\circ}$, where $\mathcal{J}^\circ$ is constructed in the same way as $\mathcal{J}$ and

$$\psi_{i,\mu}^\circ := \frac{\psi_\mu^{\square,\circ}\big((\kappa_i^\circ)^{-1}(\cdot)\big)}{\big|\det D\kappa_i^\circ\big((\kappa_i^\circ)^{-1}(\cdot)\big)\big|^{1/2}}. \tag{2.2.46}$$

By construction, Ψ° is a Riesz basis in $L_2(\Omega)$ with the same global Sobolev regularity as $\Psi^{\square,\circ}$.

It turns out that the localization property (2.2.41) is in fact fulfilled by the canonical dual of Ψ for any aggregated wavelet frame constructed in this way, as long as $\beta, \sigma > 0$ are appropriately chosen and Ψ is localized to the Riesz basis Ψ°:

Proposition 2.23. *Let Ψ and Ψ° be constructed as above and consider the function ϱ_1 from (2.2.31) for the fixed parameters $\sigma > 0$ and $0 < \beta < 2\sigma$. If, for some $\alpha > 2\max\{\frac{\sigma d}{\beta}, d\}$, it is*

$$\big|\langle \psi_\lambda, \psi_{\lambda'}^\circ \rangle\big| \lesssim \mathrm{e}^{-\alpha\varrho_2(\lambda,\lambda')}, \quad \lambda \in \mathcal{J}, \lambda' \in \mathcal{J}^\circ \tag{2.2.47}$$

where $\varrho_2 : \mathcal{J} \times \mathcal{J}^\circ \to \mathbb{R}_+^0$ is defined by

$$\begin{aligned}
\varrho_2(\lambda, \lambda') := &\frac{\beta}{\sigma} \log\Big(1 + 2^{\min\{j,j'\}}\big\|\kappa_i(2^{-j}\mathbf{k}) - \kappa_{i'}^\circ(2^{-j'}\mathbf{k}')\big\|\Big) \\
&+ |j - j'|\log 2 + \frac{9\beta}{2\sigma}\log 2,
\end{aligned} \tag{2.2.48}$$

completely analogous to (2.2.31), then there exists $\delta \in (0, \frac{\sigma}{2} - \max\{\frac{\sigma d}{\beta}, d\})$, such that

$$\big|\langle \psi_\lambda, \widetilde{\psi_{\lambda'}} \rangle\big| \lesssim \mathrm{e}^{-\delta\varrho_1(\lambda,\lambda')}, \quad \lambda, \lambda' \in \mathcal{J}. \tag{2.2.49}$$

Proof. Using an analogous proof as in Lemma 2.17, it is straightforward to see that the triple $\varrho = (\varrho_1, \varrho_2, \varrho_3)$, where

$$\begin{aligned}
\varrho_3(\lambda, \lambda') := &\frac{\beta}{\sigma} \log\Big(1 + 2^{\min\{j,j'\}}\big\|\kappa_i^\circ(2^{-j}\mathbf{k}) - \kappa_{i'}^\circ(2^{-j'}\mathbf{k}')\big\|\Big) \\
&+ |j - j'|\log 2 + \frac{9\beta}{2\sigma}\log 2,
\end{aligned}$$

is a generalized Lemarié metric for the pair of index sets $(\mathcal{J}, \mathcal{J}^\circ)$. Moreover, we can assume that a corresponding version of Lemma 2.18 also holds for ϱ, i.e., we have $\|\mathbf{E}_\gamma(\mathcal{J}, \mathcal{J}^\circ)\|_\infty < \infty$ and $\|\mathbf{E}_\gamma(\mathcal{J}^\circ, \mathcal{J}^\circ)\|_\infty < \infty$ for all $\gamma > \max\{\frac{\sigma d}{\beta}, d\}$.

Expanding the global canonical dual elements $\widetilde{\psi_\lambda}$ in the dual reference Riesz basis $\widetilde{\Psi^\circ} = \{\widetilde{\psi_\lambda^\circ}\}_{\lambda \in \mathcal{J}^\circ}$, we get

$$\langle \psi_\lambda, \widetilde{\psi_{\lambda''}} \rangle = \sum_{\lambda' \in \mathcal{J}^\circ} \langle \psi_\lambda, \psi_{\lambda'}^\circ \rangle \langle \widetilde{\psi_{\lambda'}^\circ}, \widetilde{\psi_{\lambda''}} \rangle. \tag{2.2.50}$$

By (2.2.47), it is $\left|\langle\Psi,\Psi^{\circ}\rangle^{\top}\right| \lesssim \mathbf{E}_{\alpha}(\mathcal{J},\mathcal{J}^{\circ})$, and the norms $\|\mathbf{E}_{\gamma}(\mathcal{J},\mathcal{J}^{\circ})\|_{\infty}$ and $\|\mathbf{E}_{\gamma}(\mathcal{J}^{\circ},\mathcal{J}^{\circ})\|_{\infty}$ stay bounded for all $\gamma \in (\max\{\frac{\sigma d}{\beta},d\},\alpha)$. So Theorem 2.13 implies the existence of $\gamma \in (0,\frac{\alpha}{2}-\max\{\frac{\sigma d}{\beta},d\})$ with $\left|\langle\widetilde{\Psi}^{\circ},\widetilde{\Psi}\rangle^{\top}\right| \lesssim \mathbf{E}_{\delta}(\mathcal{J}^{\circ},\mathcal{J})$. Since moreover $\alpha - \delta > \frac{\alpha}{2} + \max\{\frac{\sigma d}{\beta},d\} > \max\{\frac{\sigma d}{\beta},d\}$, we know that $\left\|\mathbf{E}_{\alpha-\delta}(\mathcal{J},\mathcal{J}^{\circ})\right\|_{\infty} < \infty$, so that Lemma 2.11 finally yields (2.2.49):

$$\left|\langle\Psi,\widetilde{\Psi}\rangle^{\top}\right| \lesssim \left|\mathbf{E}_{\alpha}(\mathcal{J},\mathcal{J}^{\circ})\mathbf{E}_{\delta}(\mathcal{J}^{\circ},\mathcal{J})\right| \lesssim \left\|\mathbf{E}_{\alpha-\delta}(\mathcal{J},\mathcal{J}^{\circ})\right\|_{\infty}\mathbf{E}_{\delta}(\mathcal{J},\mathcal{J}).$$

$\square$

It turns out that condition (2.2.47) can be fulfilled quite easily by choosing an appropriate pair of reference wavelet frames Ψ, Ψ°. Here we will exploit the fact that for all known constructions of wavelet bases and frames on the unit cube $\square$, the supports of $\psi^{\square}_{(j,\mathbf{e},\mathbf{k})}$ and $\psi^{\square,\circ}_{(j',\mathbf{e}',\mathbf{k}')}$ are essentially localized at the dyadic grid points $2^{-j}\mathbf{k}$ and $2^{-j'}\mathbf{k}'$, respectively:

$$\sup_{\mathbf{x}\in\mathrm{supp}(\psi^{\square}_{j,\mathbf{e},\mathbf{k}})} \|\mathbf{x}-2^{-j}\mathbf{k}\| \lesssim 2^{-j}, \quad (j,\mathbf{e},\mathbf{k})\in\mathcal{J}^{\square}, \qquad (2.2.51)$$

$$\sup_{\mathbf{x}\in\mathrm{supp}(\psi^{\square,\circ}_{j,\mathbf{e},\mathbf{k}})} \|\mathbf{x}-2^{-j}\mathbf{k}\| \lesssim 2^{-j}, \quad (j,\mathbf{e},\mathbf{k})\in\mathcal{J}^{\square,\circ}. \qquad (2.2.52)$$

(2.2.51) and (2.2.52) indeed hold for the constructions from [59, 61]. Since the local parametrizations κ_i and $\kappa_{i'}^{\circ}$ are sufficiently smooth, it immediately follows that also

$$\sup_{\mathbf{x}\in\mathrm{supp}\,\psi_{(i,j,\mathbf{e},\mathbf{k})}} \left\|\mathbf{x}-\kappa_i(2^{-j}\mathbf{k})\right\|_{\mathbb{R}^d} \lesssim 2^{-j}, \quad (i,j,\mathbf{e},\mathbf{k})\in\mathcal{J} \qquad (2.2.53)$$

and

$$\sup_{\mathbf{x}\in\mathrm{supp}\,\psi^{\circ}_{(i',j',\mathbf{e}',\mathbf{k}')}} \left\|\mathbf{x}-\kappa_{i'}^{\circ}(2^{-j'}\mathbf{k}')\right\|_{\mathbb{R}^d} \lesssim 2^{-j'}, \quad (i,j,\mathbf{e},\mathbf{k})\in\mathcal{J}^{\circ}. \qquad (2.2.54)$$

Then, raising some vanishing moment conditions on the reference systems $\Psi^{\square}$ and $\Psi^{\square,\circ}$ is sufficient to guarantee (2.2.47):

Theorem 2.24. *Assume that, for $N\in\mathbb{N}$ with $N\geq\alpha>0$, the systems $\Psi^{\square},\Psi^{\square,\circ}\subset H^{\alpha}(\square)$ fulfill the following moment conditions:*

$$\int_{\square} x^{\beta}\psi^{\square}_{(j,\mathbf{e},\mathbf{k})}(\mathbf{x})\,\mathrm{d}\mathbf{x} = 0, \quad |\beta|<N,\ (j,\mathbf{e},\mathbf{k})\in\mathcal{J}^{\square},\ \mathbf{e}\neq 0, \qquad (2.2.55)$$

$$\int_{\square} x^{\beta}\psi^{\square,\circ}_{(j',\mathbf{e}',\mathbf{k}')}(\mathbf{x})\,\mathrm{d}\mathbf{x} = 0, \quad |\beta|<N,\ (j',\mathbf{e}',\mathbf{k}')\in\mathcal{J}^{\square},\ \mathbf{e}'\neq 0. \qquad (2.2.56)$$

Then, lifting $\Psi^{\square}$ and $\Psi^{\square,\circ}$ as in (2.2.4) and (2.2.46), Ψ is exponentially ϱ-localized to Ψ°, i.e., there exists a constant $C>0$, only depending on global parameters, such that

$$\left|\langle\psi_{\lambda},\psi^{\circ}_{\lambda'}\rangle\right| \leq Ce^{-\alpha\varrho_2(\lambda,\lambda')}, \quad \lambda\in\mathcal{J},\lambda'\in\mathcal{J}^{\circ}, \qquad (2.2.57)$$

where ϱ_2 is given by (2.2.48), for the parameters $\sigma>0$ and $0<\beta<2\sigma$.

Proof. First of all, assume that $j' \geq j$. Using (2.2.56) and the Cauchy–Schwarz inequality, we get for $\lambda = (i, j, \mathbf{e}, \mathbf{k})$ and $\lambda' = (i', j', \mathbf{e}', \mathbf{k}')$ with $\mathbf{c}, \mathbf{c}' \neq \mathbf{0}$

$$
\begin{aligned}
\left| \langle \psi_\lambda, \psi_{\lambda'}^\circ \rangle \right| &= \left| \langle \psi_\lambda, \psi_{\lambda'}^\circ \rangle_{L_2(\text{supp}\, \psi_{\lambda'}^\circ)} \right| \\
&= \left| \int_{\text{supp}\, \psi_{(j',\mathbf{e}',\mathbf{k}')}^{\square,\circ}} \psi_\lambda(\kappa_{i'}^\circ(\mathbf{x})) \psi_{(j',\mathbf{e}',\mathbf{k}')}^{\square,\circ}(\mathbf{x}) |\det D\kappa_{i'}^\circ(\mathbf{x})|^{1/2}\, d\mathbf{x} \right| \\
&= \left| \int_{\text{supp}\, \psi_{(j',\mathbf{e}',\mathbf{k}')}^{\square,\circ}} \left(\psi_\lambda(\kappa_{i'}^\circ(\mathbf{x})) |\det D\kappa_{i'}^\circ(\mathbf{x})|^{1/2} - P(\mathbf{x}) \right) \psi_{(j',\mathbf{e}',\mathbf{k}')}^{\square,\circ}(\mathbf{x})\, d\mathbf{x} \right| \\
&\lesssim \left\| (\psi_\lambda \circ \kappa_{i'}^\circ) |\det D\kappa_{i'}^\circ(\cdot)|^{1/2} - P \right\|_{L_2(\text{supp}\, \psi_{(j',\mathbf{e}',\mathbf{k}')}^{\square,\circ})},
\end{aligned}
$$

where P is an arbitrary polynomial of total degree strictly less than N. Then a Whitney-type estimate yields

$$
\begin{aligned}
\left| \langle \psi_\lambda, \psi_{\lambda'}^\circ \rangle \right| &\lesssim 2^{-\alpha j'} \left| (\psi_\lambda \circ \kappa_{i'}^\circ) |\det D\kappa_{i'}^\circ(\cdot)|^{1/2} \right|_{H^\alpha(\text{supp}\, \psi_{(j',\mathbf{e}',\mathbf{k}')}^{\square,\circ})} \\
&\lesssim 2^{-\alpha j'} \left\| \psi_\lambda \circ \kappa_{i'}^\circ \right\|_{H^\alpha(\text{supp}\, \psi_{(j',\mathbf{e}',\mathbf{k}')}^{\square,\circ})} \\
&\lesssim 2^{-\alpha j'} \left\| \psi_\lambda \right\|_{H^\alpha(\Omega_i)} \\
&\lesssim 2^{-\alpha(j'-j)}.
\end{aligned}
$$

In the other case $j' \leq j$, one can show in a completely analogous way

$$
\begin{aligned}
\left| \langle \psi_\lambda, \psi_{\lambda'}^\circ \rangle \right| &= \left| \langle \psi_{\lambda'}^\circ, \psi_\lambda \rangle_{L_2(\text{supp}\, \psi_\lambda)} \right| \\
&= \left| \int_{\text{supp}\, \psi_{(j,\mathbf{e},\mathbf{k})}^{\square}} \psi_{\lambda'}^\circ(\kappa_i(\mathbf{x})) \psi_{(j,\mathbf{e},\mathbf{k})}^{\square}(\mathbf{x}) |\det D\kappa_i(\mathbf{x})|^{1/2}\, d\mathbf{x} \right| \\
&= \left| \int_{\text{supp}\, \psi_{(j,\mathbf{e},\mathbf{k})}^{\square}} \left(\psi_{\lambda'}^\circ(\kappa_i(\mathbf{x})) |\det D\kappa_i(\mathbf{x})|^{1/2} - P(\mathbf{x}) \right) \psi_{(j,\mathbf{e},\mathbf{k})}^{\square}(\mathbf{x})\, d\mathbf{x} \right| \\
&\lesssim \left\| (\psi_{\lambda'}^\circ \circ \kappa_i) |\det D\kappa_i(\cdot)|^{1/2} - P \right\|_{L_2(\text{supp}\, \psi_{(j,\mathbf{e},\mathbf{k})}^{\square})},
\end{aligned}
$$

so that $\left| \langle \psi_\lambda, \psi_{\lambda'}^\circ \rangle \right| \lesssim 2^{-\alpha|j-j'|}$. Now let us analyze the situations where the integrals $\langle \psi_\lambda, \psi_{\lambda'}^\circ \rangle$ can be nontrivial at all. By (2.2.53) and (2.2.54), a necessary condition for $\text{supp}\, \psi_\lambda \cap \text{supp}\, \psi_{\lambda'}^\circ$ having nontrivial measure is

$$
\left\| \kappa_i(2^{-j}\mathbf{k}) - \kappa_{i'}^\circ(2^{-j'}\mathbf{k}') \right\| \lesssim 2^{-\min\{j,j'\}},
$$

i.e.,

$$
\left(1 + 2^{\min\{j,j'\}} \left\| \kappa_i(2^{-j}\mathbf{k}) - \kappa_{i'}^\circ(2^{-j'}\mathbf{k}') \right\| \right)^{-r} \gtrsim 2^{-r}, \tag{2.2.58}
$$

for any $r > 0$ desired, where the constant involved does not depend on the concrete value of r. We choose $r := \alpha\beta/\sigma$, so that (2.2.58) yields the claim

$$
\begin{aligned}
\left| \langle \psi_\lambda, \psi_{\lambda'}^\circ \rangle \right| &\lesssim \left(1 + 2^{\min\{j,j'\}} \left\| \kappa_i(2^{-j}\mathbf{k}) - \kappa_{i'}^\circ(2^{-j'}\mathbf{k}') \right\| \right)^{-\alpha\beta/\sigma} 2^{-\alpha|j-j'|} \\
&\eqsim e^{-\alpha\varrho_2(\lambda,\lambda')}.
\end{aligned}
$$

$\square$

Part II

Discretization of Elliptic Problems

Chapter 3

Regularity Theory for Elliptic Boundary Value Problems

In the succeeding chapters, we will be interested in the numerical solution of the elliptic operator equation (0.0.11). Since the approximability of the variational solution u depends on its smoothness properties, we will collect some results from the regularity theory of second–order elliptic boundary value problems on a bounded domain $\Omega \subset \mathbb{R}^d$ in this chapter.

In the classical regularity theory, the verification of smoothness properties for u requires that the boundary $\partial\Omega$ has a sufficiently high Hölder regularity. Under the latter assumption, it can be shown that the Sobolev regularity of the unknown solution u is essentially determined by the smoothness of the right–hand side f, as the following special case of [92, Th. 9.1.16] for second–order equations shows.

Theorem 3.1. *Assume that $\Omega \in C^{1+\delta}$ for some $\delta \geq 0$ and let the bilinear form a from (0.0.4) be H^1–elliptic. Moreover, let $s \geq 0$ satisfy $s \neq \frac{1}{2}$ and $0 \leq s \leq \delta$ if $\delta \in \mathbb{N}$, and $0 \leq s < \delta$ otherwise. Concerning the coefficients $a_{\alpha,\beta}$, we assume that*

$$\partial^\gamma a_{\alpha,\beta} \in L_\infty(\Omega), \quad |\gamma| \leq \max\{0, \delta + |\beta| - 1\}$$

in the case $\delta \in \mathbb{N}$, and that

$$a_{\alpha,\beta} \in \begin{cases} C^{\delta+|\beta|-1}(\overline{\Omega}), & |\beta| > 0 \\ L_\infty(\Omega), & |\beta| = 0 \end{cases}$$

holds if $\delta \notin \mathbb{N}$. Then, for a right–hand side $f \in H^{-1+s}(\Omega)$, the variational solution u of (0.0.9) belongs to $H^{1+s}(\Omega) \cap H_0^1(\Omega)$, and it is

$$|u|_{H^{1+s}(\Omega)} \lesssim |f|_{H^{-1+s}(\Omega)} + |u|_{H^1(\Omega)}. \tag{3.0.1}$$

Consequently, for the Poisson equation (0.0.3), we may expect that u is contained in $H^{1+s}(\Omega)$ whenever $f \in H^{-1+s}(\Omega)$ and the domain is in $C^{1+s'}$ for some $s' \geq s$. Analogous smoothing results are known for convex domains Ω, where for $f \in L_2(\Omega)$ it follows that $u \in H^2(\Omega)$, see [104].

Unfortunately, the domains used in practical computations are often only piecewise smooth and have reentrant corners spoiling convexity. As an example we mention the L–shaped domain $\Omega = (-1, 1)^2 \setminus [0, 1)^2$ that shall be used in the numerical experiments of Chapters 5 and 8. In the case of such a nonsmooth domain the above mentioned classical regularity results do no longer apply. For polygonal or, more generally, for Lipschitz domains, the solution operator A^{-1} generally maps $L_2(\Omega)$ onto a larger space than $H^2(\Omega) \cap H_0^1(\Omega)$ which also comprises functions with singularities.

We shall from now on assume that the operator under consideration is the negative Dirichlet Laplacian $A = -\Delta$ induced by the bilinear form (0.0.5). Then, for the case of Lipschitz domains, the following two important regularity theorems were proved in [98].

Theorem 3.2 ($H^{3/2}$–Theorem). *Let $\Omega \subset \mathbb{R}^d$ be a bounded Lipschitz domain. If $f \in L_2(\Omega)$ and u is the weak solution of the Poisson equation (0.0.3), then $u \in H^{3/2}(\Omega)$.*

Theorem 3.3. *For each $\alpha > \frac{3}{2}$, there exists a bounded Lipschitz domain $\Omega \subset \mathbb{R}^d$ and a right–hand side $f \in C^\infty(\overline{\Omega})$, such that the weak solution u of (0.0.3) does not belong to $H^\alpha(\Omega)$.*

As a consequence of Theorem 3.2, we can infer the continuous embedding

$$D(A; L_2(\Omega)) \hookrightarrow H^{3/2}(\Omega) \cap H_0^1(\Omega), \tag{3.0.2}$$

and the solution operator

$$A^{-1} : L_2(\Omega) \to H^{3/2}(\Omega) \cap H_0^1(\Omega)$$

of the Poisson problem (0.0.3) with right–hand side $f \in L_2(\Omega)$ is bounded, see also [103, Corollary 1.25]. Conversely, Theorem 3.3 implies that the embedding (3.0.2) is sharp. Hence, for an arbitrary Lipschitz domain Ω and $f \in L_2(\Omega)$, we cannot conclude higher Sobolev regularity of the weak solution than $u \in H^{3/2}(\Omega) \cap H_0^1(\Omega)$.

Then the question arises whether u does have a higher regularity in other scales of smoothness spaces, namely, in the scale of Besov spaces. Results in this direction are referred to as *non–classical* regularity theory. In [43, 48], the following theorem was proved for the case of Lipschitz domains.

Theorem 3.4 ([43, Th. 3.1.6(a)]). *Let $\Omega \subset \mathbb{R}^d$ be a bounded Lipschitz domain. Then there exists an $\epsilon \in (0,1)$ only depending on the Lipschitz character of Ω, such that whenever u is the weak solution of the Poisson equation (0.0.3) for a right–hand side*

$$f \in B_p^{\mu-2}(L_p(\Omega)), \quad \mu \geq 1 + \frac{1}{p}, \quad 1 < p \leq 2 + \epsilon, \tag{3.0.3}$$

we know that

$$u \in B_\tau^\alpha(L_\tau(\Omega)), \quad 0 < \alpha < \min\left\{\frac{d}{d-1}\left(1 + \frac{1}{p}\right), \mu\right\}, \quad \tau = \left(\frac{\alpha}{d} + \frac{1}{p}\right)^{-1}. \tag{3.0.4}$$

In a slightly weaker form, a theorem of this type has also been proved in [48, Theorem 4.1]. By an interpolation argument, one can easily infer the following regularity result for u in the special scale of Besov spaces $B_\tau^{sd+1}(L_\tau(\Omega))$, $\tau = (s+\frac{1}{2})^{-1}$, see [45]. These particular spaces play a crucial role when it comes to nonlinear approximation in $H^1(\Omega)$, see also Section 4.1.

Corollary 3.5. *Let $\Omega \subset \mathbb{R}^d$ be a bounded Lipschitz domain and assume that $f \in H^{-1+\mu}(\Omega)$ for some $\mu \geq 1$. Then the weak solution u of the Poisson equation (0.0.3) fulfills*

$$u \in B^\alpha_\tau(L_\tau(\Omega)), \quad 1 < \alpha < \min\left\{\frac{d}{2(d-1)}, \frac{\mu+1}{3}\right\} + 1, \quad \tau = \left(\frac{\alpha-1}{d} + \frac{1}{2}\right)^{-1}. \tag{3.0.5}$$

Hence for a sufficiently smooth right–hand side f, we may expect that u has a significantly higher regularity in the scale of Besov spaces $B^{sd+1}_\tau(L_\tau(\Omega))$, $\tau = (s + \frac{1}{2})^{-1}$, than in the scale of Sobolev spaces $H^s(\Omega)$.

Remark 3.6. *It is a convenient mnemonic to visualize regularity results like Corollary 3.5 in an $s - \frac{1}{p}$ diagram, also referred to as a DeVore–Triebel diagram. Each point $(s, \frac{1}{p})^\top$ in the plane corresponds to a Besov space $B^s_q(L_p(\Omega))$. Real interpolation and embeddings between two Besov spaces then have a graphical counterpart, just by connecting the two respective points in the diagram. As an example, the Sobolev embedding theorem corresponds to a line with slope $\frac{1}{d}$, d being the space dimension of the underlying domain. A graphical "proof" of Corollary 3.5 can be found in Figure 3.1.*

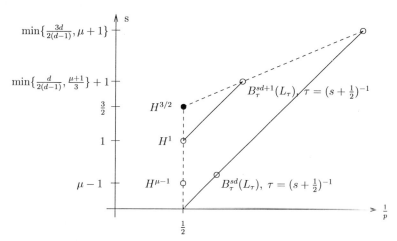

Figure 3.1: DeVore–Triebel diagram for Corollary 3.5

For the case the case that the right–hand side f is not contained in $L_2(\Omega)$, one may apply the following variant of a well–known theorem from Jerison and Kenig [98], see also [43, 45]:

Theorem 3.7. *Let $\Omega \subset \mathbb{R}^d$ be a bounded Lipschitz domain and assume that $f \in H^{\mu-1}(\Omega)$ for some $\mu > -\frac{1}{2}$. Then the weak solution u of the Poisson equation (0.0.3) fulfills*

$$u \in B^\alpha_\tau(L_\tau(\Omega)), \quad 0 < \alpha < \min\left\{\frac{3d}{2(d-1)}, \mu+1\right\}, \quad \tau = \left(\frac{\alpha}{d} + \frac{1}{2}\right)^{-1}. \tag{3.0.6}$$

There are even stronger results if the underlying domain Ω is polygonal. We will confine the discussion to the case of spatial dimension $d = 2$. Concerning the geometrical features of Ω, we shall use in the sequel the following notation as in [55, 85]. We denote the segments of $\partial\Omega$ by $\overline{\Gamma}_j$, $j = 1,\dots,N$, where Γ_j are open and numbered in positive orientation. S_j shall be the endpoint of Γ_j, with ω_j being the measure of the interior angle at S_j. We shall also need following singularity functions $S_{j,1}$, given in local polar coordinates (r_j, θ_j) in the vicinity of the corner S_j:

$$S_{j,1}(r_j, \theta_j) := \eta(r_j) r_j^{\pi/\omega_j} \sin\left(\frac{\pi\theta_j}{\omega_j}\right). \tag{3.0.7}$$

Here $\eta : \mathbb{R}_+^0 \to \mathbb{R}_+$ is a suitable cutoff function, being 1 in a neighborhood of 0 and going to zero fast enough to ensure that the supports of the N singularity functions $S_{j,1}$ do not mutually intersect.

Using this notation, regularity results for solutions of the Poisson equation in the case of polygonal domains are mainly based on the following theorem from [85].

Theorem 3.8. *Let $\Omega \subset \mathbb{R}^2$ be open, bounded and polygonal. Then, given a right–hand side $f \in L_2(\Omega)$, the variational solution u to the Poisson equation (0.0.3) decomposes into a regular part $u_R \in H^2(\Omega) \cap H_0^1(\Omega)$ and a singular part*

$$u - u_R = u_S = \sum_{\omega_j > \pi} c_j S_{j,1}, \tag{3.0.8}$$

with the singularity functions $S_{j,1}$ from (3.0.7).

By the specific decay property of $S_{j,1}$ in the vicinity of the corner S_j, it is $S_{j,1} \in H^s(\Omega)$ for $s < \min\{1 + \pi/\omega_j, 2\}$, see also [85, Th. 1.2.18]. Consequently, since we may assume to have at least one reentrant corner in a nonconvex polygonal domain Ω, u_S is contained in $H^s(\Omega)$ only for

$$s < \min\{1 + \pi/\omega_j : \omega_j > \pi\} \tag{3.0.9}$$

which may be close to $\frac{3}{2}$ if the angle ω_j of the reentrant corner is big. In contrast to the limited Sobolev regularity of u_S, it was shown in [44] that u_S has an arbitrary high regularity in a specific scale of Besov spaces.

Theorem 3.9 ([44, Th. 2.3]). *For the corner singularity functions $S_{j,1}$, it holds that $S_{j,1} \in B_\tau^\alpha\left(L_\tau(\Omega)\right)$ for all $\alpha > 0$, where $\tau = (\frac{\alpha}{2} + \frac{1}{2})^{-1}$.*

As a consequence, by interpolation between the Besov spaces $H^{3/2+\epsilon}(\Omega) = B_2^{3/2+\epsilon}(L_2(\Omega))$ and $B_\tau^\alpha(L_\tau(\Omega))$, it is $u_S \in B_\tau^\alpha(L_\tau)$ for all $\alpha > 0$ and $\tau = (\frac{\alpha-1}{2} + \frac{1}{2})^{-1}$, see [44, Theorem 2.4]. The overall Besov regularity of u for a right–hand side $f \in L_2(\Omega)$ is hence only limited by the Besov regularity of the regular part u_R.

For right–hand sides f of higher regularity than L_2, it is possible to expand the variational solution u of (0.0.3) into additional higher order singularity functions, see [85, Ch. 2.7].

Theorem 3.10. *Let $\Omega \subset \mathbb{R}^2$ be open, bounded and polygonal. Then, given a right-hand side $f \in H^{-1+s}(\Omega)$ for $s \geq 0$, the variational solution u to the Poisson equation (0.0.3) decomposes into a regular part $u_R \in H^{1+s}(\Omega) \cap H_0^1(\Omega)$ and a singular part*

$$u - u_R = u_S = \sum_{j=1}^{N} \sum_{0 < m\pi/\omega_j < s+1} c_{j,m} S_{j,m}, \tag{3.0.10}$$

where the singularity functions $S_{j,m}$ are defined as

$$S_{j,m}(r_j, \theta_j) := \begin{cases} \eta(r_j) r_j^{m\pi/\omega_j} \sin\left(\frac{m\pi\theta_j}{\omega_j}\right), & m\pi/\omega_j \notin \mathbb{Z} \\ \eta(r_j) r_j^{m\pi/\omega_j} \left(\log r_j \sin\left(\frac{m\pi\theta_j}{\omega_j}\right) + \theta_j \sin\left(\frac{m\pi\theta_j}{\omega_j}\right) \right), & otherwise \end{cases} . \tag{3.0.11}$$

Then, analogously to the situation of Theorem 3.9, also here it can be shown that u_S has limited Sobolev regularity, whereas the functions $S_{j,m}$ are in $B_\tau^\alpha(L_\tau(\Omega))$, $\tau = (\frac{\alpha}{2} + \frac{1}{2})^{-1}$, for all $\alpha > 0$, see [44].

It should be noted that the above mentioned regularity results for the Poisson equation immediately carry over to the *Helmholtz equation*

$$(\gamma I + A)u = f \text{ in } \Omega, \quad u|_{\partial\Omega} = 0, \tag{3.0.12}$$

where $\gamma > 0$ and $A = -\Delta$ is again the Dirichlet Laplacian. Problems of this type will appear in the time discretization of the heat equation in Chapter 7. In order to derive regularity estimates for the weak solution u of (3.0.12), we can apply the *resolvent equation*

$$(\lambda I - A)^{-1} - (\mu I - A)^{-1} = (\mu - \lambda)(\lambda I - A)^{-1}(\mu I - A)^{-1}, \quad \lambda, \mu \in \rho(A) \tag{3.0.13}$$

for the special case $\lambda = 0$ and $\mu = -\gamma$, where $A : D(A; V) \subset V \to V$, which gives the decomposition

$$(\gamma I + A)^{-1} = A^{-1}(I - \gamma(\gamma I + A)^{-1}). \tag{3.0.14}$$

Consequently, for a right–hand side f, the weak solution u to the Helmholtz equation (3.0.12) can then be interpreted as the weak solution of the Poisson equation with a modified right–hand side $\tilde{f} = (I - \gamma(\gamma I + A)^{-1})f$. Since $(\gamma I + A)^{-1}$ maps at least into $H^1(\Omega)$, we have that $\tilde{f} \in H^{\min\{1,s\}}(\Omega)$ whenever $f \in H^s(\Omega)$, so that for a significant range of right–hand sides, we can derive analogous Besov regularity results also for the Helmholtz equation (3.0.12).

Chapter 4

Wavelet Discretization

This chapter deals with some well–known concepts and algorithms for the numerical treatment of elliptic operator equations by means of wavelet methods.

Since we are particularly interested in the approximation of the unknown solution u up to a prescribed target accuracy $\varepsilon > 0$, we shall review the basic elements of nonlinear approximation theory in Section 4.1. It will turn out that for those operator equations we are interested in, the most important approximation methods to study are best or near–best N–term approximations. Given a Riesz basis for the energy space, we can resort to the problem of computing approximate N–term approximations in ℓ_2, which is discussed in Section 4.2. Section 4.3 provides a brief review of the fundamental properties of elliptic operators in wavelet coordinates. In the sequel, we will address the numerical realization of the three building blocks for adaptive wavelet schemes: adaptive thresholding (Section 4.4), routines for the approximation of right–hand sides (Section 4.5) and the approximate application of biinfinite compressible matrices to finite vectors (Section 4.6).

4.1   Nonlinear Approximation

Let $(X, \|\cdot\|_X)$ be a normed linear space, $v \in X$ and assume that we are dealing with the numerical approximation of v within some prescribed tolerance $\varepsilon > 0$, using only finite many basis functions from a set $\Psi - \{\psi_\lambda\}_{\lambda \in \mathcal{J}} \subset X$. It is then obvious that for decreasing tolerances ε, the number of active basis elements as well as the associated computational work and storage requirements will in general increase. Consequently, one will be interested in those algorithms where the balance between the accuracy and the associated computational cost is somewhat optimal. In the following, we shall explain in which sense optimality is meant here.

Any approximation of v will be chosen from some subspace $S \subset X$. Essentially, there are two different approximation strategies here. Either S is taken to be a linear space, e.g., the linear span

$$S_N = \operatorname{span}\{\psi_{\lambda_k}, 1 \leq k \leq N\}$$

of N wavelets. This leads to so–called *linear approximation methods*. Or we let the algorithm choose the approximations from a nonlinear set, which is referred to

71

as *nonlinear approximation*. A prominent example of a nonlinear approximation method arises if S is chosen to be the union

$$\Sigma_N = \bigcup_{\#\Lambda \leq N} S(\Lambda)$$

of all linear combinations from Ψ with at most N nontrivial coefficients, where $S(\Lambda)$ is defined as in (1.1.8). This approach is called *N–term approximation*. Obviously, Σ_N is not a linear space since the sum of two elements $x, y \in \Sigma_N$ might have $2N$ nontrivial coefficients in general. Both for linear and for nonlinear approximation approximation, we can then define the error of best approximation

$$\operatorname{dist}_X(v, S) := \inf_{w \in S} \|v - w\|_X \qquad (4.1.1)$$

and ask for which elements $v \in X$ a specific decay rate of $\operatorname{dist}_X(v, S)$ may be expected as N tends to infinity, [32, 45, 69].

More precisely, for any Banach space X and a sequence $\mathcal{T} = (T_n)_{n \geq 0}$ of nested and asymptotically dense subsets $T_n \subset X$, one introduces the *approximation space* $\mathcal{A}_q^s(X)$ related to $\mathcal{T}$ by

$$\mathcal{A}_q^s(X) := \big\{ f \in X : |f|_{\mathcal{A}_q^s(X)} < \infty \big\}, \qquad (4.1.2)$$

where $s > 0$, $0 < q \leq \infty$, and

$$|f|_{\mathcal{A}_q^s(X)} := \begin{cases} \big(\sum_{n=0}^{\infty} (n^s \operatorname{dist}_X(f, T_n))^q \frac{1}{n} \big)^{1/q} & , \ 0 < q < \infty \\ \sup_{n \geq 0} n^s \operatorname{dist}_X(f, T_n) & , \ q = \infty \end{cases} . \qquad (4.1.3)$$

$\mathcal{A}_q^s(X)$ is a Banach space for $q \geq 1$ and a quasi–Banach space for $q < 1$, with quasi–norm $\| \cdot \|_{\mathcal{A}_q^s(X)} := \| \cdot \|_X + | \cdot |_{\mathcal{A}_q^s(X)}$. Because of the monotonicity of the sequence $(\operatorname{dist}_X(f, T_n))_{n \geq 0}$, we also have the equivalence

$$|f|_{\mathcal{A}_q^s(X)} \approx \begin{cases} \big(\sum_{n=0}^{\infty} (2^{ns} \operatorname{dist}_X(f, T_{2^n}))^q \big)^{1/q} & , \ 0 < q < \infty \\ \sup_{n \geq 0} 2^{ns} \operatorname{dist}_X(f, T_{2^n}) & , \ q = \infty \end{cases} , \qquad (4.1.4)$$

which is sometimes more convenient to work with. Varying the parameters s or q, we obtain dense and continuous embeddings [70]

$$\mathcal{A}_{q_1}^{s_1}(X) \hookrightarrow \mathcal{A}_{q_2}^{s_2}(X), \quad \text{if } s_1 > s_2 \text{ or } (s_1 = s_2, \ q_1 > q_2). \qquad (4.1.5)$$

Under appropriate conditions on the sequence $\mathcal{T}$ of approximating spaces, it can be shown that $\mathcal{A}_q^s(X)$ coincides with more classical function spaces. This is possible mainly due to the following theorem which identifies $\mathcal{A}_q^s(X)$ as an interpolation space, see [32] for a proof:

Theorem 4.1. *Let $Y \hookrightarrow X$ be two densely and continuously embedded Banach spaces. Moreover, assume that $\mathcal{T} = (T_n)_{n \geq 0}$ is a sequence of nested subspaces of Y such that for some $m > 0$, we have a Jackson–type estimate*

$$\operatorname{dist}_X(f, T_n) \lesssim 2^{-mn} \|f\|_Y, \quad f \in Y \qquad (4.1.6)$$

and a Bernstein–type estimate

$$\|f\|_Y \lesssim 2^{mn}\|f\|_X, \quad f \in T_n. \tag{4.1.7}$$

Then, for $0 < s < m$, we have the norm equivalence

$$\|(2^{ns}K(f, 2^{-mn}))_{n\geq 0}\|_{\ell_q(\mathbb{N})} \approx \|f\|_X + \|(2^{ns}\operatorname{dist}_X(f, T_n))_{n\geq 0}\|_{\ell_q(\mathbb{N})}, \tag{4.1.8}$$

where $K(f, \cdot)$ is the K-functional from (1.2.4), and hence $[X, Y]_{s/m,q} = \mathcal{A}_q^s(X)$.

For the concrete situations we are interested in, namely the approximation within a Sobolev space, we can immediately draw the following two conclusions, see also [31, 32, 69]:

Corollary 4.2. *Under appropriate approximation and regularity properties of the underlying wavelet basis Ψ, we have the following facts about approximation in $X = H^t(\Omega)$:*

(i) For linear approximation in $H^t(\Omega)$, the corresponding approximation space for $q = 2$ is given as $\mathcal{A}_2^s(H^t(\Omega)) = H^{sd+t}(\Omega)$.

(ii) For N–term approximation with the spaces $T_n = \Sigma_n$, the corresponding approximation space for $q = 2$ is given as $\mathcal{A}_2^s(H^t(\Omega)) = B_\tau^{sd+t}(L_\tau(\Omega))$, where τ and s are related via $\tau = (s + \frac{1}{2})^{-1}$.

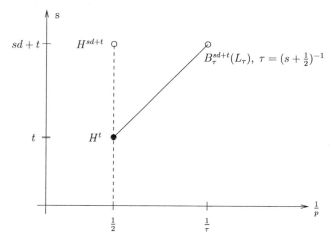

Figure 4.1: Approximation spaces in Corollary 4.2

Since any linear subspace T with $\dim T = N$ is contained in the set Σ_N, best N–term approximation can be used as the ultimate benchmark for both linear and nonlinear approximation methods. This is why we shall strive to realize at least an approximate best N–term approximations in the following. Although $H^{sd+t}(\Omega) =$

$B_2^{sd+t}(L_2(\Omega))$ is not embedded into $B_\tau^{sd+t}(L_\tau(\Omega))$ for $\tau \leq 2$, it turns out that in many cases of practical interest, the target object v may have a significantly higher smoothness in the Besov scale $B_\tau^{sd+\tau}(L_\tau)$ than in the Sobolev scale H^{sd+t} for other reasons. We have already discussed some examples in Chapter 3. In this case, nonlinear approximation method pay off most because the rate of best N–term approximation in H^t is higher than the approximation rate of, e.g., a uniform space refinement.

4.2   Best N–Term Approximation in ℓ_2

Having a Riesz basis $\Psi = \{\psi_\lambda\}_{\lambda \in \mathcal{J}}$ for the function space X at hand, any approximation of an element $v \in X$ is by definition equivalent to an approximation of the corresponding coefficient sequence $\mathbf{v}$ in ℓ_2. In this section, we therefore resort to nonlinear approximation results in ℓ_2 and show how they are related to the results of the previous section.

Let us assume that we want to approximate a given vector $\mathbf{v} = (v_\lambda)_{\lambda \in \mathcal{J}} \in \ell_2$ by another vector $\mathbf{v}_\varepsilon$ up to a given target accuracy $\varepsilon > 0$. Then the most economical such approximation would of course be the vector $\mathbf{v}_N$, defined by replacing all but the N largest coefficients in modulus of v_λ by zero, with $N = N(\varepsilon, \mathbf{v})$ being the smallest integer such that

$$\|\mathbf{v} - \mathbf{v}_N\|_{\ell_2} \leq \varepsilon. \tag{4.2.1}$$

Such a vector $\mathbf{v}_N$ is called a *best N–term approximation*, since $\mathbf{v}_N$ attains the minimal error of all N–term approximations in ℓ_2

$$\sigma_N(\mathbf{v}) := \inf \{\|\mathbf{v} - \mathbf{w}_N\|_{\ell_2} : \#\operatorname{supp} \mathbf{w}_n \leq N\}. \tag{4.2.2}$$

We are particularly interested in those subclasses of ℓ_2, where the error of best N–term approximation decays with a specific *rate $s > 0$*

$$\sigma_N(\mathbf{v}) \lesssim N^{-s}, \tag{4.2.3}$$

so that we can bound the number of significant coefficients in $\mathbf{v}_N$ by $N \lesssim \varepsilon^{-1/s}$. Essentially, these spaces correspond to the special approximation spaces $\mathcal{A}^s := \mathcal{A}_\infty^s(\ell_2)$ from (4.1.2), endowed with the equivalent (quasi–)norm

$$\|\mathbf{v}\|_{\mathcal{A}^s} := \sup_{N \geq 0}(N + 1)^s \sigma_N(\mathbf{v}), \quad \sigma_0(\mathbf{v}) := \|\mathbf{v}\|_{\ell_2}. \tag{4.2.4}$$

It turns out that the abstract set $\mathcal{A}^s$ coincides with the *weak ℓ_τ spaces* [70]

$$\ell_\tau^w := \{\mathbf{v} \in \ell_2 : |\mathbf{v}|_{\ell_\tau^w} := \sup_{n \geq 1} n^{1/\tau}|\gamma_n(\mathbf{v})| < \infty\}, \quad 0 < \tau < 2. \tag{4.2.5}$$

Here, for any $\mathbf{v} \in \ell_2$ and $n \in \mathbb{N}$, we denote by $\gamma_n(\mathbf{v})$ the n-th largest coefficient in modulus of $\mathbf{v}$. The expression $|\cdot|_{\ell_\tau^w}$ is a quasi–seminorm on ℓ_τ^w, since the triangle inequality only holds up to a τ–dependent constant

$$|\mathbf{v} + \mathbf{w}|_{\ell_\tau^w} \leq \tilde{C}_1(\tau)\big(|\mathbf{v}|_{\ell_\tau^w} + |\mathbf{w}|_{\ell_\tau^w}\big), \quad \mathbf{v}, \mathbf{w} \in \ell_\tau^w. \tag{4.2.6}$$

In addition, we set $\|\mathbf{v}\|_{\ell_\tau^w} := \|\mathbf{v}\|_{\ell_2} + |\mathbf{v}|_{\ell_\tau^w}$. The continuous and dense embeddings

$$\ell_\tau \hookrightarrow \ell_\tau^w \hookrightarrow \ell_{\tau+\delta}, \quad 0 < \delta \le 2 - \tau, \tag{4.2.7}$$

are the reason for which ℓ_τ^w is called weak ℓ_τ space. The connection between $\mathcal{A}^s$ and ℓ_τ^w is clarified in the following theorem from [33, 69]:

Theorem 4.3. *For $s > 0$ and $\tau = (s + \frac{1}{2})^{-1}$, the spaces $\mathcal{A}^s$ and ℓ_τ^w coincide with equivalent norms $\|\cdot\|_{\mathcal{A}^s} \approx \|\cdot\|_{\ell_\tau^w}$:*

$$|\mathbf{v}|_{\ell_\tau^w} \approx \sup_{N \ge 1} N^s \sigma_N(\mathbf{v}), \quad \mathbf{v} \in \ell_\tau^w. \tag{4.2.8}$$

Especially, $\mathbf{v} \in \ell_\tau^w$ implies that $\sigma_N(\mathbf{v}) \le CN^{-s}$, where the constant C only depends on τ as τ tends to zero.

The following example shows how well the so far stated results fit together when it comes to nonlinear approximation in a Sobolev space and an appropriate wavelet Riesz basis is used:

Example 4.4. *Assume that H is a closed subspace of $H^t(\Omega)$ and that $v \in H$ shall be approximated with a best N–term approximation in $H^t(\Omega)$. Moreover, assume that $\{\Psi, \tilde{\Psi}\}$ is a wavelet Riesz basis in $L_2(\Omega)$ such that $\mathbf{D}^{-t}\Psi$ is also a Riesz basis in H. Let us denote by $\mathbf{v} := \mathbf{D}^{-t}\langle v, \tilde{\Psi}\rangle^\top \in \ell_2$ the coefficient sequence of v with respect to $\mathbf{D}^{-t}\Psi$. From Corollary 4.2 we know that the approximation space for best N–term approximation in $H^t(\Omega)$ is $\mathcal{A}_2^s(H^t(\Omega)) = B_\tau^{sd+t}(L_\tau(\Omega))$ with $\tau = (s + \frac{1}{2})^{-1}$, as long as the underlying approximating spaces fulfill the appropriate direct and inverse estimates. For the particularly interesting special case of spline wavelet bases with approximation order $m \ge sd + t$, this is indeed the case, see [33, 55, 134].*

Then, by (1.2.24), we have that $v \in B_\tau^{sd+t}(L_\tau(\Omega))$ is equivalent to $\mathbf{v} \in \ell_\tau$. Since $\ell_\tau \hookrightarrow \ell_\tau^w = \mathcal{A}^s$ by (4.2.7) and Theorem 4.3, we finally get the implication

$$v \in B_\tau^{sd+t}(L_\tau(\Omega)), \ \tau = \left(s + \frac{1}{2}\right)^{-1} \quad \Rightarrow \quad \sigma_N(\mathbf{v}) \lesssim N^{-s}, \quad N \in \mathbb{N}. \tag{4.2.9}$$

In practical computations, where the target quantity $\mathbf{v}$ is only implicitly given, we cannot expect the exact rearrangement $\gamma_N(\mathbf{v})$ to be known. Then it is easier to aim at the development of algorithms that compute an *approximate* or *near* best N–term approximation $\mathbf{w}_N$, by which we mean that $\# \operatorname{supp} \mathbf{w}_N \le N$ and

$$\|\mathbf{v} - \mathbf{w}_N\|_{\ell_2} \le C\sigma_N(\mathbf{v}), \tag{4.2.10}$$

where $C \ge 1$ is a uniform constant. In the estimates for the computational complexity of such an algorithm, the best N–term approximation $\mathbf{v}_N$ may then serve as a benchmark. To explain this in more detail, note that the approximation of an element $\mathbf{v} \in \mathcal{A}^s = \ell_\tau^w$ up to accuracy ε needs at most $N \approx \varepsilon^{-1/s}$ degrees of freedom. Hence we get the lower bound $\varepsilon^{-1/s}$ for the computational complexity to determine any $\mathbf{v}$ from the unit ball from $\mathcal{A}^s$ within accuracy ε. Therefore it is natural to call an algorithm *asymptotically optimal* if it realizes $\varepsilon^{-1/s}$ also as an upper complexity bound, see [35]. More precisely, we call an algorithm s^*–*optimal*, if for any $\mathbf{v}$ from the unit ball in $\mathcal{A}^s$, $s < s^*$, and a given target accuracy ε, the algorithm yields an approximation $\mathbf{v}_\varepsilon$ with $\|\mathbf{v} - \mathbf{v}_\varepsilon\|_{\ell_2} \le \varepsilon$ and the number of nontrivial coefficients in $\mathbf{v}_\varepsilon$ as well as the number of arithmetic operations and storage locations to compute $\mathbf{v}_\varepsilon$ stay proportional to $\varepsilon^{-1/s}$.

4.3   Elliptic Operators in Wavelet Coordinates

For the numerical solution of the original operator equation (0.0.11), we shall of course utilize some wavelet Riesz basis $D^{-t}\Psi$ in the energy space H as ansatz and test functions. Doing so, it is well–known that the original operator equation (0.0.11) can be reformulated as an equivalent biinfinite matrix equation

$$Au = f,   \qquad (4.3.1)$$

where $u = u^\top D^{-t}\Psi$, $f = D^{-t}\langle f, \Psi\rangle^\top$ and $A = D^{-t}\langle A\Psi, \Psi\rangle^\top D^{-t}$ is the diagonally preconditioned stiffness matrix. This guiding principle has already been observed and propagated in a variety of early papers in wavelet theory, see, e.g., [13, 57, 60].

As a composition of the operator A and the Riesz maps, $A : \ell_2 \to \ell_2$ is boundedly invertible, which in turn implies the existence of constants $c_1, c_2 \geq 0$, such that

$$c_1\|v\|_{\ell_2} \leq \|Av\|_{\ell_2} \leq c_2\|v\|_{\ell_2}, \quad v \in \ell_2. \qquad (4.3.2)$$

Consequently, the spectral condition number of A is bounded by

$$\kappa(A) = \|A\|_{\mathcal{L}(\ell_2)}\|A^{-1}\|_{\mathcal{L}(\ell_2)} \leq c_2 c_1^{-1}. \qquad (4.3.3)$$

In the case of $A = (a_{\lambda,\lambda'})_{\lambda,\lambda' \in \mathcal{J}}$ being positive definite, results of the type (4.3.3) carry over also to submatrices $A_\Lambda := (a_{\lambda,\lambda'})_{\lambda,\lambda' \in \Lambda}$, with a set $\Lambda \subset \mathcal{J}$ of active wavelet coefficients. Here we have the estimates

$$\|A_\Lambda\|_{\mathcal{L}(\ell_2(\Lambda))} \leq \|A\|_{\mathcal{L}(\ell_2)}, \quad \|A_\Lambda^{-1}\|_{\mathcal{L}(\ell_2(\Lambda))} \leq \|A^{-1}\|_{\mathcal{L}(\ell_2)}. \qquad (4.3.4)$$

Consequently, also the condition numbers of the submatrices A_Λ stay uniformly bounded by

$$\kappa(A_\Lambda) \leq \kappa(A) \leq c_2 c_1^{-1}. \qquad (4.3.5)$$

In the sequel, we shall need the *discrete energy norm*

$$\|v\|_A := \langle v, Av\rangle_{\ell_2} \qquad (4.3.6)$$

which, in view on the assumptions on A is also equivalent to the H–norm:

$$\|v\|_H \approx \|v\|_a \approx \|v\|_{\ell_2} \approx \|v\|_A, \quad v = v^\top D^{-t}\Psi. \qquad (4.3.7)$$

4.4   Building Block 1: Adaptive Thresholding

In the numerical algorithms of the succeeding chapters, we will have to compute best or near best N–term approximations of a given *finitely* supported vector v in linear time, i.e., with at most a multiple of $\#\operatorname{supp} v$ operations. Since the computation of the exact best N–term approximation v_N would require the sorting all elements of v by their modulus, such a direct approach would need asymptotically a constant times $\#\operatorname{supp} v \log(\#\operatorname{supp} v)$ arithmetic operations and therefore precludes itself. However, in [6, 133] it has been shown that a complete sorting of the entries of v is not necessary when it suffices to compute an approximate best N–term approximation $\tilde{v}_N$, i.e., $\|v - \tilde{v}_N\|_{\ell_2} \leq C\varepsilon$. Using approximate sorting techniques like binary binning or bucket sort algorithms [42], the additional log-factor in the complexity estimate can in fact be avoided. There, one uses the alternative characterization of ℓ_τ^w

$$\ell_\tau^w = \left\{ \mathbf{v} \in \ell_2 : \#\{\lambda : 2^{-j} \geq |v_\lambda| > 2^{-(j+1)}\} \lesssim 2^{j\tau}, \, j \in \mathbb{Z} \right\} \tag{4.4.1}$$

and one aims at regrouping the entries of $\mathbf{v}$ according to the corresponding equivalence classes (*bins*) of dyadic orders of magnitude

$$V_i = \left\{ (\lambda, v_\lambda) : 2^{-(i+1)} < \frac{|v_\lambda|}{\|\mathbf{v}\|_{\ell_2}} \leq 2^{-i} \right\}. \tag{4.4.2}$$

This approximate sorting can be performed in linear time. Reformulated as a numerical procedure **COARSE**, the computation of an approximate best N–term approximation for a given finitely supported $\mathbf{v}$ looks as follows, see also [133]:

Algorithm 4.5. COARSE$[\mathbf{v}, \varepsilon] \to \mathbf{v}_\varepsilon$:

- $q := \lceil \log((\# \operatorname{supp} \mathbf{v})^{1/2} \|\mathbf{v}\|_{\ell_2} / \varepsilon) \rceil$

- *Regroup the elements of $\mathbf{v}$ into the sets $V_0, \ldots, V_q$, where $v_\lambda \in V_i$ if and only if $2^{-(i+1)} \|\mathbf{v}\|_{\ell_2} < |v_\lambda| \leq 2^{-i} \|\mathbf{v}\|_{\ell_2}$, $0 \leq i < q$. Possible remaining elements are put into the set V_q.*

- *Create $\mathbf{v}_\varepsilon$ by successively extracting elements from V_0 and when it is empty from V_1 and so forth, until $\|\mathbf{v} - \mathbf{v}_\varepsilon\|_{\ell_2} \leq \varepsilon$.*

It should be noted that it is possible to implement the **COARSE** algorithm without actually constructing the bins V_i, see [6]. Of course, this does only affect the constant in the complexity estimate. The following properties of **COARSE** can be shown [133]:

Proposition 4.6. *Let $\mathbf{v}$ be finitely supported and $\mathbf{v}_\varepsilon := \mathbf{COARSE}[\mathbf{v}, \varepsilon]$. Then we have $\|\mathbf{v} - \mathbf{v}_\varepsilon\|_{\ell_2} \leq \varepsilon$ and $\mathbf{v}_\varepsilon$ has at most $\# \operatorname{supp} \mathbf{v}_\varepsilon \lesssim \inf\{N : \sigma_N(\mathbf{v}) \leq \varepsilon\}$ significant entries. Moreover, the number of arithmetic operations and storage locations needed to compute $\mathbf{v}_\varepsilon$ is bounded by a constant multiple of $\# \operatorname{supp} \mathbf{v} + \max\{\log(\varepsilon^{-1} \|\mathbf{v}\|_{\ell_2}), 1\} \lesssim \varepsilon^{-1/s} |\mathbf{v}|_{\ell_\tau^w}^{1/s}$.*

The routine **COARSE** is frequently used in adaptive algorithms in order to ensure asymptotic optimality. This is mainly due to the following fact [33, 133]:

Proposition 4.7. *Let $\theta < 1/3$ be fixed, $\tau \in (0, 2)$ and $\tau = (s + \frac{1}{2})^{-1}$. Then, for any $\varepsilon > 0$, $\mathbf{v} \in \ell_\tau^w$, and a finitely supported approximation $\mathbf{w} \in \ell_2$ with $\|\mathbf{v} - \mathbf{w}\|_{\ell_2} \leq \theta \varepsilon$, the output $\overline{\mathbf{w}} := \mathbf{COARSE}[\mathbf{w}, (1 - \theta)\varepsilon]$ fulfills $\|\mathbf{v} - \overline{\mathbf{w}}\|_{\ell_2} \leq \varepsilon$ and the number of significant entries in $\overline{\mathbf{w}}$ is bounded by*

$$\# \operatorname{supp} \overline{\mathbf{w}} \lesssim \varepsilon^{-1/s} |\mathbf{v}|_{\ell_\tau^w}^{1/s}. \tag{4.4.3}$$

As a consequence, there is a constant $\tilde{C}_2(\tau)$, only depending on τ, such that

$$|\overline{\mathbf{w}}|_{\ell_\tau^w} \leq \tilde{C}_2(\tau) |\mathbf{v}|_{\ell_\tau^w}. \tag{4.4.4}$$

However, it should be noted that there are so far also adaptive wavelet algorithms the convergence of which does not require any coarsening of the iterands, see [82].

For the proof of many complexity estimates for adaptive algorithms in ℓ_2, including Proposition 4.7 and the results presented in Chapter 7, one needs the following important perturbation result for sequences from ℓ_τ^w. The proof can be found in [133].

Lemma 4.8 ([33, Lemma 4.11],[133, Proposition 3.4]). *Let* $\tau \in (0, 2)$ *with* $\tau = (s + 1/2)^{-1}$. *Then, for* $\mathbf{v} \in \ell_\tau^w$ *and any finitely supported* $\mathbf{z} \in \ell_2$, *we have the estimate*

$$|\mathbf{z}|_{\ell_\tau^w} \lesssim |\mathbf{v}|_{\ell_\tau^w} + (\# \operatorname{supp} \mathbf{z})^s \|\mathbf{v} - \mathbf{z}\|_{\ell_2}. \tag{4.4.5}$$

4.5　Building Block 2: Approximate Input Data

For the design of an adaptive scheme, one always has to assume that the input data that determine the problem under consideration are accessible in a specific sense. Concerning the class of elliptic operator equations (0.0.11) we are interested in, we shall assume that the right–hand side $f \in H'$ or, equivalently, its infinite wavelet expansion coefficients $\mathbf{f} = \mathbf{D}^{-t}\langle f, \Psi \rangle^\top$ are completely known. By this we mean that we are able to compute approximate wavelet expansions of f in the dual basis up to any given accuracy. More strictly speaking, we require that for any $\varepsilon > 0$, there exists a computable, finitely supported array $\mathbf{f}_\varepsilon \in \ell_2(\mathcal{J})$ such that

$$\|\mathbf{f} - \mathbf{f}_\varepsilon\|_{\ell_2} \leq \varepsilon. \tag{4.5.1}$$

Since we know that the dual wavelet basis $\mathbf{D}^t\tilde{\Psi}$ is dense in H', this first assumption is not critical. However, in view of the pending complexity analysis of the adaptive algorithm, it is of course also necessary that the computation of an approximate right–hand side is done in the most economical way. To this end, we shall firstly consider only those exact right–hand sides $\mathbf{f}$ that are contained in some Lorentz sequence space ℓ_τ^w, $\tau = (s + \frac{1}{2})^{-1}$, $s < s^*$. By the wavelet characterization results from Section 1.2, this assumption is equivalent to saying that f has a specific Besov regularity.

Concerning the approximate right–hand sides, we will then require that $\mathbf{f}_\varepsilon$ realize approximate best N–term approximations of the exact infinite right–hand side $\mathbf{f}$, i.e.,

$$\# \operatorname{supp} \mathbf{f}_\varepsilon \lesssim |\mathbf{f}|_{\ell_\tau^w}^{1/s} \varepsilon^{-1/s}. \tag{4.5.2}$$

Moreover, the associated computational work should stay proportional to the size $\# \operatorname{supp} \mathbf{f}_\varepsilon$ of the input data. In the sequel, the computation of approximate right–hand sides will be referred to as the numerical subroutine

$$\mathbf{RHS}[t, \varepsilon] \to \mathbf{f}_\varepsilon. \tag{4.5.3}$$

The assumptions on **RHS** can be justified in practice by using a priori information on the singular and the smooth parts of the right–hand side f. A possible realization may also consist of a projection of f onto a fine multiresolution space $\tilde{V}_j$, followed by a thresholding step with the **COARSE** routine from Section 4.4.

4.6 Building Block 3: Adaptive Matrix–Vector Multiplication

4.6.1 Decay Estimates

In order to design an adaptive wavelet scheme for the approximate solution of (4.3.1), it turns out to be of crucial importance that in many cases, the entries of the biinfinite system matrix $\mathbf{A}$ exhibit a fast off–diagonal decay. Therefore, just by dropping matrix entries that are small in modulus, $\mathbf{A}$ may be approximated well by a sparse matrix with only a finite number of entries per row and column.

As the most important example, consider the case where H is a closed subspace of the Sobolev space $H^t(\Omega)$. There, for a large class of local and non–local elliptic operators $A : H \to H'$, the stiffness matrix of an appropriate wavelet discretization exhibits decay estimates of the form

$$2^{-(|\lambda'|+|\lambda|)t}|\langle A\psi_{\lambda'},\psi_\lambda\rangle| \lesssim 2^{-||\lambda|-|\lambda'||\sigma}(1+\delta(\lambda,\lambda'))^{-\beta}, \quad \lambda,\lambda' \in \mathcal{J}, \qquad (4.6.1)$$

where $\sigma > d/2$, $\beta > d$ and δ is given by (2.2.12).

Estimates of the form (4.6.1) are known to hold as soon as the underlying primal wavelet basis Ψ is sufficiently smooth and the wavelets ψ_λ have adequate cancellation properties [33, 129]. In particular, let us mention the discussion in [55, 134]. There, under the assumptions that A is bounded from H^{t+s} to H^{-t+s} for $|s| \leq \tau$ and that Ψ admits the characterization (1.2.16) of Sobolev spaces H^s for $s \in (-\tilde{\gamma}, \gamma)$, it was shown that (4.6.1) holds for $\sigma = \min\{\tau, \gamma - t, t + \tilde{m}\}$. The value of β is greater or equal to $d + 2\tilde{m} + 2t$ in the case of integral operators, and β can even be chosen arbitrarily large in the case of differential operators.

In view of the estimate (4.6.1), we introduce for parameters $\sigma, \beta > 0$ the class of matrices $\mathcal{A}_{\sigma,\beta} = \mathcal{A}_{\sigma,\beta,\delta}$, comprising all matrices $\mathbf{B} = (b_{\lambda,\lambda'})_{\lambda,\lambda'\in\mathcal{J}}$, such that

$$|b_{\lambda,\lambda'}| \leq c_{\mathbf{B}} 2^{-||\lambda|-|\lambda'||\sigma}(1+\delta(\lambda,\lambda'))^{-\beta}, \quad \lambda,\lambda' \in \mathcal{J} \qquad (4.6.2)$$

holds for a constant $c_{\mathbf{B}}$ which only depends on $\mathbf{B}$. A matrix $\mathbf{B}$ is called *quasi–sparse*, if it is $\mathbf{B} \in \mathcal{A}_{\sigma,\beta}$ for some $\sigma > d/2$ and $\beta > d$. By an application of the Schur Lemma, it can be shown that any quasi–sparse matrix $\mathbf{B}$ is bounded on ℓ_2, which has not been explicitly required in the definition. We refer to [33, Prop. 3.3] for a proof.

Moreover, as it turns out, quasi–sparse matrices are especially interesting from the computational point of view, since they can be approximated well by sparse matrices with only finite many nonzero entries per row and per column. To explain this in more detail, we call a bounded operator $\mathbf{A} \in \mathcal{L}(\ell_2)$ s^*–*compressible*, when for each $j \in \mathbb{N}$ there exists an infinite matrix $\mathbf{A}_j$ with at most $\alpha_j 2^j$ nontrivial entries in each row and column with $\sum_{j\in\mathbb{N}} \alpha_j < \infty$, such that for any $s < s^*$, we have $\|\mathbf{A} - \mathbf{A}_j\|_{\mathcal{L}(\ell_2)} \leq C_j$ and $\sum_{j\in\mathbb{N}} C_j 2^{js}$ is summable. Whenever one of these matrices $\mathbf{A}_j$ is applied to a vector with at most $n = 2^j$ significant entries, then the corresponding number of arithmetical operations is bounded by a constant multiple of n. More generally, it can be shown that a given s^*–compressible matrix $\mathbf{B}$ maps ℓ_τ^w boundedly onto itself, where $\tau = (s + \frac{1}{2})^{-1}$ and $0 \leq s < s^*$, see [33, Prop. 3.8] for a proof.

It remains to verify whether and for which concrete value of s^* a given quasi–sparse stiffness matrix $\mathbf{A}$ from the wavelet discretization (4.3.1) can be expected to be s^*–compressible. This question has been addressed and answered positively in a series of papers [7, 33, 47, 55, 129, 134], at various levels of generality. Especially for the case of spline wavelet bases, the attainable value for s^* turns out to be significantly high. The most comprehensive results in this direction can be found in [134], where for spline wavelets of order m and a large class of differential and integral operators of order t, s^*–compressibility of $\mathbf{A}$ could be established for a value $s^* > \frac{m-t}{d}$. This lower bound is of particular importance in view of Example 4.4, since $\frac{m-t}{d}$ is exactly the maximal convergence rate of the best N–term approximation with these wavelet bases, only assuming that $u \in B_\tau^{sd+t}(L_\tau(\Omega))$. Having this in mind, it is desirable not to spoil the decay properties of the current iterand by approximate multiplications with $\mathbf{A}$ that are unavoidable in nearly every adaptive wavelet algorithm.

Remark 4.9. *By the concrete construction given in Section 2.2, analogous compressibility results immediately transfer to the case of aggregated wavelet frames. In fact, the smooth lifting of the reference wavelet basis on $\square = (0,1)^d$ to the domain Ω preserves both the cancellation properties and the local regularity of the frame elements. Consequently, the building blocks of adaptive wavelet schemes, as discussed in this chapter, are also available for discretizations based on spline wavelet frames.*

4.6.2   Approximate Application of Compressible Matrices

Having an s^*–compressible matrix $\mathbf{A}$ at hand one will of course be interested in algorithms that realize the approximate application of $\mathbf{A}$ to finitely supported vectors $\mathbf{v}$ within a given target accuracy, where the associated computational work stays proportional to the length of the input parameter. In [33], it has been possible to design a numerical routine **APPLY** the output of which approximates the exact matrix–vector product $\mathbf{A}\mathbf{v}$ with the desired tolerance and that has linear computational complexity, up to sorting operations. Analogous to the already mentioned routine **COARSE**, binning and approximate sorting strategies may be used to eliminate these sorting costs and to obtain an asymptotically optimal algorithm. Consequently, we shall work with the following refined variant of **APPLY** from [133]:

Algorithm 4.10. APPLY$[\mathbf{A}, \mathbf{v}, \varepsilon] \to \mathbf{w}_\varepsilon$:

- $q := \lceil \log((\# \operatorname{supp} \mathbf{v})^{1/2} \|\mathbf{v}\|_{\ell_2} \|\mathbf{A}\|_{\mathcal{L}(\ell_2)} 2/\varepsilon) \rceil$

- *Regroup the elements of $\mathbf{v}$ into the sets $V_0, \ldots, V_q$, where $v_\lambda \in V_i$ if and only if $2^{-(i+1)} \|\mathbf{v}\|_{\ell_2} < |v_\lambda| \le 2^{-i} \|\mathbf{v}\|_{\ell_2}$, $0 \le i < q$. Possible remaining elements are put into the set V_q.*

- *For $k = 0, 1, \ldots$, generate vectors $\mathbf{v}_{[k]}$ by subsequently extracting $2^k - \lfloor 2^{k-1} \rfloor$ elements from $\bigcup_i V_i$, starting from V_0 and when it is empty continuing with V_1 and so forth, until for some $k = l$ either $\bigcup_i V_i$ becomes empty or*

$$\|\mathbf{A}\|_{\mathcal{L}(\ell_2)} \left\| \mathbf{v} - \sum_{k=0}^{l} \mathbf{v}_{[k]} \right\|_{\ell_2} \le \varepsilon/2. \qquad (4.6.3)$$

In both cases, $\mathbf{v}_{[l]}$ *may contain less than* $2^l - \lfloor 2^{l-1} \rfloor$ *elements.*

- *Compute the smallest* $j \geq l$ *such that*

$$\sum_{k=0}^{l} C_{j-k} \|\mathbf{v}_{[k]}\|_{\ell_2} \leq \varepsilon/2. \tag{4.6.4}$$

- *For* $0 \leq k \leq l$, *compute the non–zero entries in the matrices* $\mathbf{A}_{j-k}$ *which have a column index in common with one of the entries of* $\mathbf{v}_{[k]}$ *and compute*

$$\mathbf{w}_\varepsilon := \sum_{k=0}^{l} \mathbf{A}_{j-k} \mathbf{v}_{[k]}. \tag{4.6.5}$$

The asymptotic optimality of **APPLY** has been shown in [133, Prop. 3.8]:

Proposition 4.11. *Let* $\mathbf{A}$ *be* s^**–compressible and* $\mathbf{v}$ *be finitely supported. Then, for the output* $\mathbf{w}_\varepsilon := \mathbf{APPLY}[\mathbf{A}, \mathbf{v}, \varepsilon]$, *we have* $\|\mathbf{A}\mathbf{v} - \mathbf{w}_\varepsilon\|_{\ell_2} \leq \varepsilon$ *and* $\mathbf{w}_\varepsilon$ *has at most* $\#\operatorname{supp}\mathbf{w}_\varepsilon \lesssim \varepsilon^{-1/s} |\mathbf{v}|_{\ell_\tau^w}^{1/s}$ *significant entries. Moreover, the number of arithmetic operations and storage locations needed to compute* $\mathbf{w}_\varepsilon$ *is bounded by a constant multiple of* $\varepsilon^{-1/s} |\mathbf{v}|_{\ell_\tau^w}^{1/s} + \#\operatorname{supp}\mathbf{v}$.

4.7 Examples

In this section, we briefly address those adaptive algorithms that shall be used in the numerical examples. The design of most adaptive wavelet algorithms follows a general paradigm from [33, 34] which essentially comprises the following steps:

1. Using an appropriate wavelet Riesz basis, reformulate the original operator equation (0.0.11) as an equivalent problem over some sequence space ℓ_2.

2. Establish a convergent approximation scheme in the full space ℓ_2 that works with infinite vectors, the exact right–hand side $\mathbf{f}$ and exact matrix–vector multiplications.

3. Then, derive an *implementable* variant of this algorithm, replacing all infinite–dimensional quantities by finitely supported and computable ones. In particular, one has to work with an inexact right–hand side $\bar{\mathbf{f}} \approx \mathbf{f}$, approximate matrix–vector operations, and appropriately matched tolerances for the subroutines like **APPLY**, **RHS** or **COARSE**. The algorithm should provide an approximation of the unknown solution up to a given target accuracy ε. Many known algorithms contain an outer loop over a geometrically decreasing sequence of tolerances $\varepsilon^{(i)} \to \varepsilon$ which facilitates the convergence and complexity analysis, but this is not necessary in general.

4. Finally, if possible, establish an optimal convergence rate of this implementable algorithm and give complexity estimates. Preferably, the convergence rate should match the rate of the best N–term approximation, and the associated computational work should behave at most linearly in the number of unknowns.

In the sequel, we will abbreviate adaptive solvers of the aforementioned type by the subroutine **SOLVE**. The defining properties of **SOLVE** shall be collected in a generic theorem:

Theorem 4.12. *Let* **B** *be* s^**-compressible and assume that for* $s \in (0, s^*)$, $\tau = (s + 1/2)^{-1}$, *the system*

$$\mathbf{Bx} = \mathbf{y}, \tag{4.7.1}$$

(4.7.1) has a solution $\mathbf{x} \in \ell_\tau^w$. *Then the numerical routine* **SOLVE**$[\mathbf{B}, \mathbf{y}, \varepsilon] \to \mathbf{x}_\varepsilon$ *produces a finitely supported* $\mathbf{x}_\varepsilon$ *so that*

$$\|\mathbf{x} - \mathbf{x}_\varepsilon\|_{\ell_2} \le \varepsilon. \tag{4.7.2}$$

Moreover, the number of nontrivial entries in $\mathbf{x}_\varepsilon$ *is bounded by*

$$\# \operatorname{supp} \mathbf{x}_\varepsilon \le C |\mathbf{x}|_{\ell_\tau^w}^{1/s} \varepsilon^{-1/s}, \tag{4.7.3}$$

and the number of arithmetic operations to compute $\mathbf{x}_\varepsilon$ *is also at most a multiple of* $\# \operatorname{supp} \mathbf{x}_\varepsilon$.

4.7.1   A Richardson Iteration

The most straightforward method to establish an ℓ_2–convergent numerical scheme for the operator equation (4.3.1) is a *Richardson iteration*

$$\mathbf{u}^{(0)} := \mathbf{0}, \quad \mathbf{u}^{(n+1)} := \mathbf{u}^{(n)} + \omega(\mathbf{f} - \mathbf{A}\mathbf{u}^{(n)}), \quad n = 0, 1, \dots \tag{4.7.4}$$

with a relaxation parameter $\omega \in \mathbb{R}$. For the convergence of (4.7.4), we require $\| \cdot \|$ to be an equivalent norm on ℓ_2 such that for the associated operator norm $\|\mathbf{M}\| := \sup_{\|\mathbf{v}\|=1} \|\mathbf{Mv}\|$, it holds that

$$\rho := \|\mathbf{I} - \omega\mathbf{A}\| < 1. \tag{4.7.5}$$

It is a well–known fact from the theory of iterative methods that under (4.7.5), the iteration (4.7.4) will exhibit linear convergence in ℓ_2 to the exact solution $\mathbf{u}$ with an error reduction per step by the factor ρ

$$\|\mathbf{u}^{(n+1)} - \mathbf{u}\| \le \rho \|\mathbf{u}^{(n)} - \mathbf{u}\|. \tag{4.7.6}$$

Moreover, in case of convergence, the exact solution $\mathbf{u}$ has the Neumann series representation

$$\mathbf{u} = \omega(\omega\mathbf{A})^{-1}\mathbf{f} = \omega \sum_{n=0}^{\infty} (\mathbf{I} - \omega\mathbf{A})^n \mathbf{f}. \tag{4.7.7}$$

For the special case of $\mathbf{A}$ being symmetric and positive definite with extremal eigenvalues $0 < \lambda_{\min} \le \lambda_{\max} = \|\mathbf{A}\|_{\mathcal{L}(\ell_2)}$, a sufficient criterion for (4.7.5) to hold is that

$$0 < \omega < \frac{2}{\lambda_{\max}}. \tag{4.7.8}$$

The optimal relaxation parameter $\hat{\omega}$ which minimizes the error reduction factor $\rho(\omega) = \|\mathbf{I} - \omega\mathbf{A}\|_{\mathcal{L}(\ell_2)}$ can be computed as

$$\hat{\omega} = \frac{2}{\lambda_{\min} + \lambda_{\max}} \qquad (4.7.9)$$

with

$$\rho(\hat{\omega}) = \frac{\lambda_{\max} - \lambda_{\min}}{\lambda_{\max} + \lambda_{\min}} = \frac{\kappa(\mathbf{A}) - 1}{\kappa(\mathbf{A}) + 1}. \qquad (4.7.10)$$

However, it should be emphasized that in order to get convergence of the Richardson iteration (4.7.4), the system matrix $\mathbf{A}$ neither has to be symmetric nor definite.

In [34], for the case of a discretization based on wavelet Riesz bases, an implementable variant of the Richardson iteration (4.7.4) was studied. It could be shown in loc. cit. that the approximate iteration converges with optimal order and that the associated computational work stays proportional to the support sizes of the iterands. Since in Chapter 5, we shall discuss a variant of this Richardson iteration that also works in the case of a frame discretization, we will omit further details on the numerical realization of (4.7.4) here.

4.7.2 The CDD1 Algorithm

In the numerical treatment of parabolic problems in Chapter 7, we will make use of the algorithm from [33], which requires the matrix $\mathbf{A}$ to be symmetric and positive definite. The algorithm introduced in loc. cit. is based on *Galerkin approximations* u_Λ, where for a given finite index set $\Lambda \subset \mathcal{J}$, one defines $u_\Lambda \in S_\Lambda$ to be the solution of the projected problem

$$a(u_\Lambda, v) = \langle f, v \rangle_{H' \times H}, \quad v \in S_\Lambda. \qquad (4.7.11)$$

By the Céa lemma [92], the Galerkin approximation u_Λ is quasi–optimal, since we have

$$\|u - u_\Lambda\|_H \leq C \inf_{v \in S_\Lambda} \|u - v\|_H \qquad (4.7.12)$$

with a constant C only depending on Ω. Hence u_Λ converges to u as soon as the spaces S_Λ are chosen arbitrarily dense in H.

Given a target accuracy $\varepsilon > 0$, the primary task of any adaptive wavelet–Galerkin scheme is then to compute a finite index set $\Lambda_\varepsilon \subset \mathcal{J}$, such that the target accuracy is realized

$$\|u - u_{\Lambda_\varepsilon}\|_H \leq \varepsilon. \qquad (4.7.13)$$

Most probably, an algorithm for the computation of Λ_ε will be of iterative form. Given an initial guess $\Lambda_0 \subset \mathcal{J}$, e.g., $\Lambda_0 = \emptyset$, the iteration loop decomposes into the following three basic steps:

1. Compute the Galerkin approximation u_{Λ_j}.

2. Estimate the error $\|u - u_{\Lambda_j}\|$ in some norm, using reliable a posteriori error estimators.

3. If necessary, refine the set of active wavelet coefficients $\Lambda_j \rightarrow \Lambda_{j+1}$, increase the iteration depth j and continue with step 1.

Of course steps 2 and 3 are the most intricate ones. In [46], it was shown how to realize a reliable and implementable a posteriori error estimator based on a wavelet expansion of the current *residual*

$$r_\Lambda := f - Au_\Lambda = A(u - u_\Lambda) \in H'. \qquad (4.7.14)$$

Since $A : H \rightarrow H'$ is boundedly invertible, the current error in $\| \cdot \|_H$ is equivalent to the H'–norm of the residual

$$\|u - u_\Lambda\|_H \approx \|r_\Lambda\|_{H'}. \qquad (4.7.15)$$

Consequently, given a wavelet Riesz basis $\mathbf{D}^t\tilde{\Psi}$ for H', we can plug in a wavelet expansion of the residual

$$r_\Lambda = \sum_{\lambda \in \mathcal{J}} 2^{-t|\lambda|} \langle r_\Lambda, \psi_\lambda \rangle \tilde{\psi}_\lambda, \qquad (4.7.16)$$

which gives the norm equivalence

$$\|u - u_\Lambda\|_H \approx \left(\sum_{\lambda \in \mathcal{J}\setminus\Lambda} 2^{-2t|\lambda|} |\langle r_\Lambda, \psi_\lambda \rangle|^2 \right)^{1/2}. \qquad (4.7.17)$$

Here we have used that by Galerkin orthogonality, those wavelet coordinates of r_Λ that refer to Λ are zero and can therefore be neglected. As a consequence of (4.7.17), the quantities

$$\delta_\lambda := 2^{-t|\lambda|} |\langle r_\Lambda, \psi_\lambda \rangle| = 2^{-t|\lambda|} \left| \langle f, \psi_\lambda \rangle - \sum_{\lambda' \in \mathcal{J}\setminus\Lambda} \langle A\psi_\lambda, \psi_{\lambda'} \rangle \right| \qquad (4.7.18)$$

may serve as *error indicators* where to refine the set Λ. Note that the series in (4.7.18) is still infinite, so that δ_λ has yet to be replaced by a computable quantity. However, it has been shown in [46] that the wavelet coefficients of r_Λ decay sufficiently fast to allow for such an approximation.

Moreover, one still has to guarantee convergence of the scheme, which concerns step 3 of the aforementioned iteration. Given an index set Λ and a Galerkin approximation u_Λ, the question is how to find a new index set $\hat{\Lambda}$, such that the error decreases geometrically in the discrete energy norm

$$\|\mathbf{u}_{\hat{\Lambda}} - \mathbf{u}\|_\mathbf{A} \leq q\|\mathbf{u}_\Lambda - \mathbf{u}\|_\mathbf{A}, \qquad (4.7.19)$$

which, due to the symmetry of A, is equivalent to the saturation property

$$\|\mathbf{u}_{\hat{\Lambda}} - \mathbf{u}_\Lambda\|_\mathbf{A} \geq \sqrt{1 - q^2}\|\mathbf{u}_\Lambda - \mathbf{u}\|_\mathbf{A}. \qquad (4.7.20)$$

But since the latter norm is equivalent to the norm $\|\mathbf{r}_\Lambda\|_{\ell_2}$ of the discrete residual, where $r_\Lambda = \mathbf{r}_\Lambda^\top \mathbf{D}^t\tilde{\Psi}$, it suffices that the new index set $\hat{\Lambda}$ covers the most significant entries of $\mathbf{r}_\Lambda$. In [46], it has been shown how to compute an approximate residual

$\mathbf{r}_\Lambda|_{\hat{\Lambda}}$ having its nontrivial entries contained in a simultaneously computed index set $\hat{\Lambda}$ and

$$\|\mathbf{r}_\Lambda|_{\hat{\Lambda}}\|_{\ell_2} \geq \eta \|\mathbf{r}_\Lambda\|_{\ell_2} \tag{4.7.21}$$

for a given parameter $\eta \in (0,1)$. Plugging the computation of an approximate residual into step 3 of the global iteration, the saturation property (4.7.20) and hence the convergence of the overall scheme could be verified in [46].

Unfortunately, it is not guaranteed by this argument that the growth *rate* of the chosen index sets Λ_j in [46] is the most economic one. In view of the results on best N–term approximation from the previous section, the ultimate goal would be to realize the same convergence order as the best N–term approximation. More precisely, one will be interested in such a Galerkin scheme that realizes the estimate

$$\|\mathbf{u} - \mathbf{u}_{\Lambda_j}\| \leq C\|\mathbf{u}\|_{\ell_\tau^w}^{1/s}(\#\Lambda_j)^{-s}, \tag{4.7.22}$$

as j tends to infinity. Essentially, one has to control the growth rate $\#\Lambda_{j+1}/\#\Lambda_j$ from one index set Λ_j to the next finer one. In [33], it was shown how the iteration from [46] can be made optimal, just by interleaving the refinement steps with coarsening steps that ensure the approximation rate of the best N–term approximation. The concrete algorithm in loc. cit. uses the procedure **COARSE** that was introduced in Section 4.2.

First numerical tests of the CDD1 algorithm were initially published in [8] which is also the basis for the concrete implementation employed in this thesis. Experiments concerning a simplified version of the scheme have been published in [39].

Chapter 5

Frame Discretization

This chapter is concerned with the design of adaptive frame algorithms for the numerical solution of the operator equation (0.0.11), where we will work essentially with the setting introduced in Chapter 3. The results presented in this Chapter are based on the analysis in [51].

Inspired by the techniques from [33], in Section 5.1 we give the principal argument that the original operator equation can be again reformulated as a matrix equation in the sequence space ℓ_2. A series representation of the exact solution immediately induces a Richardson iteration for the discrete system which shall be discussed in Section 5.2. As the exact algorithm involves infinite matrices and vectors, we derive an implementable version thereof and prove its convergence and optimal computational complexity under some technical assumptions. For the verification of the asserted convergence and complexity results, we shall give some numerical examples in Section 5.3.

5.1 Principal Ideas

By a construction as laid out in Chapter 2, we may assume from now on that we have a Gelfand frame $\Psi = \{\psi_\lambda\}$ for the Gelfand triple $(H, L_2(\Omega), H')$, with a corresponding Gelfand triple of sequence spaces $(\ell_{2,\mathbf{D}^t}, \ell_2, \ell_{2,\mathbf{D}^{-t}})$. Consequently, we know that the sequence spaces are isomorphic with isomorphisms $\varphi_{\mathbf{D}^t}$ and $\varphi_{\mathbf{D}^t}^*$ from (2.1.19) and (2.1.20), respectively. With a slight abuse of notation, we can therefore abbreviate both mappings $\varphi_{\mathbf{D}^t}$ and $\varphi_{\mathbf{D}^t}^*$ by $\mathbf{D}^t$ in the sequel, and their inverses are given by $\mathbf{D}^{-t}$. A visualization of the situation can be found in Figure 5.1.

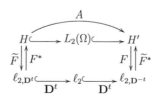

Figure 5.1: Mappings in a Gelfand frame discretization of $Au = f$.

Then it is easily shown that the original operator equation can be reformulated as an equivalent biinfinite system in frame coordinates.

Lemma 5.1. *Under the assumptions on A, the operator*

$$\mathbf{A} := \mathbf{D}^{-t} F A F^* \mathbf{D}^{-t} = \mathbf{D}^{-t} \langle A\Psi, \Psi \rangle^{\top} \mathbf{D}^{-t} \tag{5.1.1}$$

is a bounded operator from ℓ_2 to ℓ_2. Moreover $\mathbf{A} = \mathbf{A}^*$ and it is boundedly invertible on its range $\mathrm{Ran}(\mathbf{A}) = \mathrm{Ran}(\mathbf{D}^{-t}F)$.

Proof. Since $\mathbf{A}$ is a composition of bounded operators $\mathbf{D}^{-t} : \ell_2 \to \ell_{2,\mathbf{D}^t}$, $F^* : \ell_{2,\mathbf{D}^t} \to H$ $A : H \to H$, $F : H' \to \ell_{2,\mathbf{D}^{-t}}$ and $\mathbf{D}^{-t} : \ell_{2,\mathbf{D}^{-t}} \to \ell_2$, $\mathbf{A}$ is a bounded operator from ℓ_2 to ℓ_2. Moreover, from the decomposition (5.1.1) it is clear that

$$\mathrm{Ker}(\mathbf{A}) = \mathrm{Ker}(F^* \mathbf{D}^{-t}), \quad \mathrm{Ran}(\mathbf{A}) = \mathrm{Ran}(\mathbf{D}^{-t}F). \tag{5.1.2}$$

Since A is symmetric, we have $\mathbf{A} = \mathbf{A}^*$ and the orthogonal decomposition

$$\ell_2 = \mathrm{Ker}(F^* \mathbf{D}^{-t}) \oplus \mathrm{Ran}(\mathbf{D}^{-t}F). \tag{5.1.3}$$

Therefore

$$\mathbf{A}_{|\,\mathrm{Ran}(\mathbf{A})} : \ \mathrm{Ran}(\mathbf{A}) \to \mathrm{Ran}(\mathbf{A}) \tag{5.1.4}$$

is boundedly invertible.   □

In analogy to the result (4.7.7), we can again derive a series representation of the unique solution $\mathbf{Q}u$ of $\mathbf{A}u = \mathbf{f}$ in $\mathrm{Ran}(\mathbf{A})$:

Theorem 5.2. *Let A satisfy the ellipticity assumptions. Denote*

$$\mathbf{f} := \mathbf{D}^{-t} F f \tag{5.1.5}$$

and $\mathbf{A}$ as in (5.1.1). Then the solution u of (0.0.11) can be computed by

$$u = F^* \mathbf{D}^{-t} \mathbf{Q}u \tag{5.1.6}$$

where $\mathbf{u}$ solves

$$\mathbf{Q}u = \omega \sum_{n=0}^{\infty} (\mathbf{I} - \omega \mathbf{A})^n \mathbf{f}, \tag{5.1.7}$$

with $0 < \omega < 2/\lambda_{\max}$ and $\lambda_{\max} = \|\mathbf{A}\|_{\mathcal{L}(\ell_2)}$. Here $\mathbf{Q} : \ell_2 \to \mathrm{Ran}(\mathbf{A})$ is the orthogonal projection onto $\mathrm{Ran}(\mathbf{A})$.

Proof. Like in Proposition 2.2, we have the expansion $u = \sum_{\lambda \in \mathcal{J}} \langle u, \tilde{\psi}_\lambda \rangle_V \psi_\lambda$ in V. Since Ψ is a Gelfand frame, $F^* \tilde{F} : H \to H$ is bounded and implies $u = F^* \tilde{F}u = \sum_{\lambda \in \mathcal{J}} \langle u, \tilde{\psi}_\lambda \rangle_{H \times H'} \psi_\lambda$ with convergence in H. By Proposition 2.2 and the Banach frame properties of Ψ, (0.0.11) is equivalent to the system of equations

$$\sum_{\lambda' \in \mathcal{J}} \langle u, 2^{t|\lambda'|} \tilde{\psi}_{\lambda'} \rangle_{H \times H'} 2^{-t(|\lambda'|+|\lambda|)} \langle A\psi_{\lambda'}, \psi_\lambda \rangle_{H' \times H} = 2^{-t|\lambda|} \langle f, \psi_\lambda \rangle_{H' \times H}, \quad \lambda \in \mathcal{J}. \tag{5.1.8}$$

Denote $\mathbf{u} := \mathbf{D}^t \tilde{F} u$ and $\mathbf{f}$, $\mathbf{A}$ as in (5.1.5) and (5.1.1), respectively. Then (5.1.8) can be rewritten as

$$\mathbf{A}\mathbf{u} = \mathbf{f}. \tag{5.1.9}$$

For all $\mathbf{v} \in \ell_2$

$$\langle \mathbf{A}\mathbf{v}, \mathbf{v} \rangle_{\ell_2} = \langle \mathbf{D}^{-t} F A F^* \mathbf{D}^{-t} \mathbf{v}, \mathbf{v} \rangle_{\ell_2} = \langle A F^* \mathbf{D}^{-t} \mathbf{v}, F^* \mathbf{D}^{-t} \mathbf{v} \rangle_{H' \times H}.$$

Therefore, since A is positive, $\mathbf{A}$ is positive semi–definite. Let us denote $\lambda_{\max} := \|\mathbf{A}\|_{\mathcal{L}(\ell_2)}$ and $\lambda_{\min}^+ := \|(\mathbf{A}_{|\operatorname{Ran}(\mathbf{A})})^{-1}\|_{\mathcal{L}(\ell_2)}^{-1}$. For $0 < \omega < 2/\lambda_{\max}$, one can consider the operator

$$\mathbf{B} := \omega \sum_{n=0}^{\infty} (\mathbf{I} - \omega \mathbf{A})^n. \tag{5.1.10}$$

Since $\rho := \|\mathbf{I} - \omega \mathbf{A}_{|\operatorname{Ran}(\mathbf{A})}\|_{\mathcal{L}(\ell_2)} = \max\{\omega \lambda_{\max} - 1, 1 - \omega \lambda_{\min}^+\} < 1$, with minimum at $\omega^* = 2/(\lambda_{\max} + \lambda_{\min}^+)$, one has that $\mathbf{B}$ is a well–defined bounded operator on $\operatorname{Ran}(\mathbf{A})$. Moreover, it is also clear that

$$\mathbf{B} \circ \mathbf{A}_{|\operatorname{Ran}(\mathbf{A})} = \mathbf{A} \circ \mathbf{B}_{|\operatorname{Ran}(\mathbf{A})} = \operatorname{id}_{\operatorname{Ran}(\mathbf{A})}. \tag{5.1.11}$$

Since $\mathbf{A}(\mathbf{I} - \mathbf{Q}) = 0$,

$$\mathbf{A}\mathbf{u} = \mathbf{A}\mathbf{Q}\mathbf{u} = \mathbf{f}. \tag{5.1.12}$$

Therefore $\mathbf{Q}\mathbf{u} \in \operatorname{Ran}(\mathbf{A})$ is the unique solution of (5.1.9) in $\operatorname{Ran}(\mathbf{A})$ and we infer from (5.1.11) that

$$\mathbf{Q}\mathbf{u} = \mathbf{B}\mathbf{f}. \tag{5.1.13}$$

By construction

$$\begin{aligned}
\langle f, \psi_\lambda \rangle_{H' \times H} &= \langle \tilde{F}^* F f, \psi_\lambda \rangle_{H' \times H} \\
&= \langle \tilde{F}^* \mathbf{D}^t \mathbf{f}, \psi_\lambda \rangle_{H' \times H} \\
&= \langle \tilde{F}^* \mathbf{D}^t \mathbf{A}\mathbf{Q}\mathbf{u}, \psi_\lambda \rangle_{H' \times H} \\
&= \langle A F^* \mathbf{D}^{-t} \mathbf{Q}\mathbf{u}, \psi_\lambda \rangle_{H' \times H}, \quad \lambda \in \mathcal{J},
\end{aligned}$$

so that $u - F^* \mathbf{D}^{-t} \mathbf{Q}\mathbf{u}$ solves (0.0.11) $\qquad\square$

5.2 An Approximate Richardson Iteration

5.2.1 The Algorithm

As a consequence of the series representation (5.1.7), the exact Richardson iteration

$$\mathbf{u}^{(0)} := 0, \quad \mathbf{u}^{(n+1)} := \mathbf{u}^{(n)} + \omega(\mathbf{f} - \mathbf{A}\mathbf{u}^{(n)}), \quad n = 0, 1, \dots \tag{5.2.1}$$

will converge in ℓ_2 to the vector $\mathbf{Q}\mathbf{u}$, as long as the relaxation parameter is appropriately chosen according to $0 < \omega < 2/\lambda_{\max}$. It is noteable that by $\mathbf{f}, \mathbf{u}^{(0)} \in \operatorname{Ran}(\mathbf{A})$, the exact iterands $\mathbf{u}^{(n)}$ are also contained in $\operatorname{Ran}(\mathbf{A})$ and hence ℓ_2–orthogonal to $\operatorname{Ker}(\mathbf{A})$.

Of course, the Richardson iteration (5.2.1) cannot be implemented from infinite vectors. Instead, we have to replace the ingredients by finite–dimensional approximations. We require the existence of the numerical routines **COARSE** and **APPLY**, as introduced in Sections 4.4 and 4.6. Moreover, for the handling of approximate right–hand sides, we assume that we have access to a third numerical routine **RHS**$[\mathbf{f}, \varepsilon] \to \mathbf{f}_\varepsilon$, as specified in Section 4.5. As mentioned in Remark 4.9, these assumptions on the numerical subroutines can indeed be assumed to hold for appropriate aggregated spline wavelet frames.

Having these three numerical routines at hand, it is straightforward to derive the following algorithm from the ideal iteration (5.2.1):

Algorithm 5.3. SOLVE$[\mathbf{A}, \mathbf{f}, \varepsilon] \to \mathbf{u}_\varepsilon$:
Let $\theta < 1/3$ and $K \in \mathbb{N}$ be fixed such that $3\rho^K < \theta$.
$i := 0$, $\mathbf{v}^{(0)} := 0$, $\varepsilon_0 := \|(\mathbf{A}_{|\operatorname{Ran}(\mathbf{A})})^{-1}\|_{\mathcal{L}(\ell_2)}\|\mathbf{f}\|_{\ell_2}$
While $\varepsilon_i > \varepsilon$ do
    $i := i + 1$
    $\varepsilon_i := 3\rho^K \varepsilon_{i-1}/\theta$
    $\mathbf{f}^{(i)} := \mathbf{RHS}[\mathbf{f}, \frac{\theta \varepsilon_i}{6\omega K}]$
    $\mathbf{v}^{(i,0)} := \mathbf{v}^{(i-1)}$
    For $j = 1, ..., K$ do
        $\mathbf{v}^{(i,j)} := \mathbf{v}^{(i,j-1)} - \omega(\mathbf{APPLY}[\mathbf{A}, \mathbf{v}^{(i,j-1)}, \frac{\theta \varepsilon_i}{6\omega K}] - \mathbf{f}^{(i)})$
    od
    $\mathbf{v}^{(i)} := \mathbf{COARSE}[\mathbf{v}^{(i,K)}, (1 - \theta)\varepsilon_i]$
od
$\mathbf{u}_\varepsilon := \mathbf{v}^{(i)}$.

Note that here, deviating somewhat from the notation in (5.2.1), we denote by $\mathbf{v}^{(i)}$ the result after applying K approximate Richardson iterations at a time to $\mathbf{v}^{(i-1)}$.

5.2.2   Convergence and Complexity Analysis of SOLVE

Concerning the convergence of the frame algorithm **SOLVE**, one can show the following theorem, see [51, 133]:

Theorem 5.4. *In the situation of Theorem 5.2, let $\mathbf{u} \in \ell_2$ be a solution of* (5.1.9). *Then* **SOLVE**$[\mathbf{A}, \mathbf{f}, \varepsilon]$ *produces finitely supported vectors* $\mathbf{v}^{(i,K)}, \mathbf{v}^{(i)}$ *such that*

$$\left\|\mathbf{Q}(\mathbf{u} - \mathbf{v}^{(i)})\right\|_{\ell_2} \leq \varepsilon_i, \quad i \geq 0. \tag{5.2.2}$$

In particular, one has

$$\|u - F^* \mathbf{D}^{-t} \mathbf{u}_\varepsilon\|_H \leq \|F^*\|_{\mathcal{L}(\ell_{2,\mathbf{D}^t}, H)}\varepsilon. \tag{5.2.3}$$

Moreover, it holds that

$$\left\|\mathbf{Q}\mathbf{u} - (\operatorname{id} - \mathbf{Q})\mathbf{v}^{(i-1)} - \mathbf{v}^{(i,K)}\right\|_{\ell_2} \leq \frac{2\theta \varepsilon_i}{3}, \quad i \geq 1. \tag{5.2.4}$$

Proof. The proof is essentially analogous to that of [133, Prop. 2.1]. For $i = 0$, (5.1.7) and (5.1.13) yield

$$\left\|Q(u - v^{(0)})\right\|_{\ell_2} = \|Qu\|_{\ell_2} = \|Bf\|_{\ell_2} \leq \varepsilon_0.$$

Now take $i \geq 1$ and let $\left\|Q(u - v^{(i-1)})\right\|_{\ell_2} \leq \varepsilon_{i-1}$ hold. We show (5.2.4) first. When exactly performing one damped Richardson iteration (5.2.1), from, say, some vector $w^{(i)}$ to $w^{(i+1)}$, equations (5.1.9) and (5.1.12) yield

$$Qu - w^{(i+1)} = Qu - w^{(i)} + \omega(Aw^{(i)} - f) = (I - \omega A)(Qu - w^{(i)}). \tag{5.2.5}$$

So, by induction, the exact application of K damped Richardson iterations at a time would result in

$$Qu - w^{(i+K)} = (I - \omega A)^K(Qu - w^{(i)}). \tag{5.2.6}$$

But the loop of Algorithm 5.3 performs K *perturbed* Richardson iterations $v^{(i,j)}$, starting from $v^{(i,0)} = v^{(i-1)}$. By construction, the stepwise error does not exceed

$$\left\|v^{(i,j)} - v^{(i,j-1)} + \omega(Av^{(i,j-1)} - f)\right\|_{\ell_2} \leq \omega\left(\frac{\theta\varepsilon_i}{6\omega K} + \frac{\theta\varepsilon_i}{6\omega K}\right) = \frac{\theta\varepsilon_i}{3K},$$

so that after K steps we end up in

$$\left\|Qu - v^{(i,K)} - (I - \omega A)^K(Qu - v^{(i-1)})\right\|_{\ell_2} \leq K\frac{\theta\varepsilon_i}{3K} = \frac{\theta\varepsilon_i}{3}. \tag{5.2.7}$$

It is straightforward to compute the identity

$$(I - \omega A)^K(Qu - v^{(i-1)}) = (I - \omega A)^K Q(u - v^{(i-1)}) - (I - Q)v^{(i-1)}. \tag{5.2.8}$$

But due to the specific choice of the relaxation parameter ω in Theorem 5.2, we get

$$\left\|(I - \omega A)^K Q(u - v^{(i-1)})\right\|_{\ell_2} \leq \rho^K\left\|Q(u - v^{(i-1)})\right\|_{\ell_2} \leq \rho^K\varepsilon_{i-1} = \frac{\theta\varepsilon_i}{3}, \tag{5.2.9}$$

which, together with (5.2.8), yields (5.2.4). Now, by using (5.2.4) and the definition of **COARSE**, one has

$$\begin{aligned}
\left\|Qu + (I - Q)v^{(i-1)} - v^{(i)}\right\|_{\ell_2} &\leq \left\|Qu + (I - Q)v^{(i-1)} - v^{(i,K)}\right\|_{\ell_2} \\
&\quad + \|v^{(i,K)} - v^{(i)}\|_{\ell_2} \leq \frac{2\theta}{3}\varepsilon_i + (1 - \theta)\varepsilon_i \leq \varepsilon_i.
\end{aligned}$$

Then (5.2.2) follows by

$$\begin{aligned}
\|Q(u - v^{(i)})\|_{\ell_2}^2 &\leq \|Q(u - v^{(i)})\|_{\ell_2}^2 + \|(I - Q)(v^{(i-1)} - v^{(i)})\|_{\ell_2}^2 \\
&= \|Q(u - v^{(i)}) + (I - Q)(v^{(i-1)} - v^{(i)})\|_{\ell_2}^2 \\
&= \|Qu + (I - Q)v^{(i-1)} - v^{(i)}\|_{\ell_2}^2.
\end{aligned}$$

Since $\text{Ker}(F^*D^{-t}) = \text{Ker}(A) = \text{Ker}(Q)$, we finally verify

$$\begin{aligned}
\|u - F^*D^{-t}u_\varepsilon\|_H &= \left\|F^*D^{-t}(Qu - u_\varepsilon)\right\|_H \\
&= \left\|F^*D^{-t}Q(u - u_\varepsilon)\right\|_H \\
&\leq \|F^*\|_{\mathcal{L}(\ell_{2,D^t},H)}\left\|Q(u - u_\varepsilon)\right\|_{\ell_2} \\
&\leq \|F^*\|_{\mathcal{L}(\ell_{2,D^t},H)}\varepsilon.
\end{aligned}$$

$\square$

Remark 5.5. *In the proof of Theorem 5.4, contrary to [51, Th. 4.3], we have explicitly used that the isomorphism* $\mathbf{D}^t : \ell_{2,\mathbf{D}^t} \to \ell_2$ *and its* ℓ_2*-adjoint have operator norm 1. This simplification can be made in all cases of practical interest.*

Since Algorithm 5.3 essentially follows the ideas from the case of Riesz bases [34], the question is whether one can show optimality also for frame discretizations. For this purpose, one has to analyze the support sizes of the approximate iterands $\mathbf{u}^{(n)}$ in more detail. Ideally, in order to match the convergence rate s of the best N-term approximation, $\#\operatorname{supp}\mathbf{u}^{(n)}$ should grow at most like $\varepsilon^{-1/s}|\mathbf{u}|_{\ell_\tau^w}^{1/s}$, $\tau = (s + \frac{1}{2})^{-1}$, as ε tends to zero.

The delicate point for the complexity analysis of frame discretizations is that one can no longer expect the sequence of iterands $\mathbf{u}^{(n)}$ to stay bounded in ℓ_τ^w. This is essentially due to the fact that the inexact applications of $\mathbf{A}$ and the approximate right–hand sides used in the algorithm cause the iterands $\mathbf{u}^{(n)}$ to leave the space $\operatorname{Ran}(\mathbf{A})$ and to have a significant share in $\operatorname{Ker}(\mathbf{A})$. During the iteration, these kernel components in $\mathbf{u}^{(n)}$ might not be damped out by subsequent Richardson iterations, so that the errors are not summable in ℓ_τ^w. Of course, the kernel components in $\mathbf{u}^{(n)}$ are only visible as additional degrees of freedom in the discrete version of the algorithm. Any application of F^*, e.g., in a postprocessing step, will cancel them out.

To cope with this situation, additional conditions on the underlying wavelet frame seem to be inevitable at the moment. In [133], sufficient conditions were given that guarantee optimality of Algorithm 5.3. To state the corresponding result, recall the constants $\tilde{C}_1(\tau)$ and $\tilde{C}_2(\tau)$ from (4.2.6) and (4.4.4), respectively.

Theorem 5.6 ([133, Th. 3.12]). *For some* $s^* > 0$, *assume that* $\mathbf{A}$ *is* s^**-compressible,* $\mathbf{f}$ *is* s^**-optimal, and that for some* $s \in (0, s^*)$, $\tau = (s + \frac{1}{2})^{-1}$, *the system* $\mathbf{Au} = \mathbf{f}$ *has a solution* $\mathbf{u} \in \ell_\tau^w$. *Moreover, assume that there exists an* $\check{s} \in (s, s^*)$ *such that with* $\check{\tau} = (\check{s} + \frac{1}{2})^{-1}$, *the operator* $\mathbf{Q}$ *is bounded on* $\ell_{\check{\tau}}^w$. *Then, if the parameter* K *in* **SOLVE** *is sufficiently large, e.g.,*

$$3\rho^K < \theta \min\left\{1, (\tilde{C}_1(\check{\tau})\tilde{C}_2(\check{\tau})|\mathbf{I} - \mathbf{Q}|_{\mathcal{L}(\ell_{(\check{\tau})}^w)})^{s/(\check{s}-s)}\right\}, \qquad (5.2.10)$$

then for all $\varepsilon > 0$, *the output* $\mathbf{u}_\varepsilon :=$ **SOLVE**$[\mathbf{A}, \mathbf{f}, \varepsilon]$ *has at most* $\#\operatorname{supp}\mathbf{u}_\varepsilon \lesssim \varepsilon^{-1/s}|\mathbf{u}|_{\ell_\tau^w}^{1/s}$ *nontrivial entries and the number of arithmetic operations needed to compute* $\mathbf{u}_\varepsilon$ *stays bounded by at most a multiple of* $\#\operatorname{supp}\mathbf{u}_\varepsilon$.

Some remarks concerning the assumptions in Theorem 5.6 are in order here. First of all, it is required that the discrete operator equation (5.1.9) has a solution $\mathbf{u} \in \ell_\tau^w$. If $\mathbf{D}^{-t}\Psi$ is a Riesz basis in H, then regularity estimates of the form $u \in B_\tau^{sd+t}(L_\tau(\Omega))$ with $\tau = (s + \frac{1}{2})^{-1}$ are equivalent to $\mathbf{u} = \langle u, \mathbf{D}^t\tilde{\Psi}\rangle^\top \in \ell_\tau$, where $\mathbf{D}^t\tilde{\Psi}$ is the dual Riesz basis in H'. As a consequence, we immediately get $\mathbf{u} \in \ell_\tau^w$ without further assumptions on the wavelet basis.

In the case of frames, the existence of solutions $\mathbf{u} \in \ell_\tau^w$ is nontrivial. In general, the decay of the expansion coefficients $\langle u, \tilde{\psi}_\lambda\rangle$ with respect to the canonical dual $\tilde{\Psi}$ will only be sufficiently fast if $\tilde{\Psi}$ is smooth and if it has an appropriate number of vanishing moments. For wavelet frames over a bounded domain Ω, unfortunately, regularity and cancellation properties of the canonical dual are in general unknown.

In order to verify decay properties of *some* expansion coefficient array $\mathbf{u}$, it is convenient to resort to non–canonical dual frames $\Xi = \{\xi_\lambda\}_{\lambda \in \mathcal{J}}$ the properties of which are explicitly known. In the shift–invariant setting, the problem of characterizing function spaces by decay properties of non–canonical dual frame coefficients has been studied recently in [22, 23, 75]. Moreover, this very strategy is followed in [50], see also [133, Sec. 4.4], where for the special case of $H = H_0^1(\Omega)$ over the L–shaped domain

$$\Omega = (-1,1)^2 \setminus [0,1)^2 = (-1,1) \times (-1,0) \cup (-1,0) \times (-1,1), \qquad (5.2.11)$$

an appropriate aggregated wavelet frame $\Psi = \Psi^{(1)} \cup \Psi^{(2)}$ is considered. In this situation, a non–canonical dual frame is easily given by weighting the local canonical duals $\tilde{\Psi}^{(i)}$ with the partition of unity $\phi_1 = \phi \circ \theta$ and $\phi_2 = 1 - \phi_1$ from part (ii) of Remark 2.22,

$$\xi_{(i,\mu)} := \phi_i \tilde{\psi}_{(i,\mu)}, \quad (i,\mu) \in \mathcal{J}, \qquad (5.2.12)$$

since any $v \in L_2(\Omega)$ can be decomposed into

$$v = \phi_1 v + \phi_2 v = \sum_{\lambda = (i,\mu) \in \mathcal{J}} \langle \phi_i v, \tilde{\psi}_\lambda \rangle \psi_\lambda = \sum_{\lambda = (i,\mu) \in \mathcal{J}} \langle v, \phi_i \tilde{\psi}_\lambda \rangle \psi_\lambda = \sum_{\lambda \in \mathcal{J}} \langle v, \xi_\lambda \rangle \psi_\lambda.$$

Moreover, it has been shown in [52] that for any $u \in H_0^1(\Omega)$, it is $\phi_i u \in H_0^1(\Omega_i)$ with $\|\phi_i u\|_{H^1(\Omega)} \approx \|u\|_{H^1(\Omega)}$. Hence one may expect that the cancellation properties of the local canonical dual systems $\tilde{\Psi}^{(i)}$ imply the desired decay properties of the non–canonical expansion coefficients. In this context, let us only summarize some special results from [50] in the following lemma:

Lemma 5.7. *Assume, as above, that Ω is the L–shaped domain (5.2.11) and that Ψ is a wavelet Gelfand frame for $(H_0^1(\Omega), L_2(\Omega), H^{-1}(\Omega))$. Moreover, let $u = S_{1,1}$ be the singularity function (3.0.7) belonging to the reentrant corner at the origin. Then it is $\phi_i u \in B_\tau^{2s+1}(L_\tau(\Omega))$ for all $s > 0$, $\tau = (s + \frac{1}{2})^{-1}$. As a consequence, there exists $s^* > 0$, only depending on the properties of the template wavelet basis $\Psi^\square$ and its dual $\tilde{\Psi}^\square$, such that for any $s \in (0, s^*)$, the expansion coefficient array $\mathbf{u} := \langle u, \mathbf{D}\Xi \rangle_{H \times H'}^\top$ is contained in ℓ_τ^w, $\tau = (s + \frac{1}{2})^{-1}$.*

The second technical assumption in Theorem 5.6 is the boundedness of the projector $\mathbf{Q}$ on some ℓ_τ^w. For the special case that $H = L_2(\Omega)$, the wavelet bases making up the aggregated frame are $L_2(\Omega)$-orthonormal, and some damping is applied to the wavelets near the interior boundaries, it could be shown in [133] that $\mathbf{Q}$ is indeed bounded on ℓ_τ^w. However, the general proof of the boundedness assumption is a difficult open problem. Its validity can only indirectly verified by the results of numerical experiments, see Section 5.3. According to [133, Remark 3.13], the boundedness of $\mathbf{Q}$ on ℓ_τ^w for all $s \in (0, s^*)$ is almost a necessary requirement for the scheme to behave optimally.

Remark 5.8. *In the case of time–frequency localized Gabor frames, the boundedness assumption on $\mathbf{Q}$ could be rigorously proved in Theorem 7.1 of [51], see also [49] for the construction of optimal adaptive algorithms in the superordinate class of polynomially localized frames.*

We conclude this section with a final remark on possible variants of the discussed Richardson iteration.

Remark 5.9. *Since in the case of frames, a wrong choice of the relaxation parameter* α *potentially spoils the convergence and/or optimality of the overall Richardson iteration, it is reasonable to investigate alternative iterative schemes. A straightforward generalization of* (5.2.1) *is the* steepest descent scheme

$$\mathbf{u}^{(0)} = \mathbf{0}, \quad \mathbf{u}^{(n+1)} = \mathbf{u}^{(n)} + \frac{\langle \mathbf{r}^{(n)}, \mathbf{r}^{(n)} \rangle}{\langle \mathbf{r}^{(n)}, \mathbf{A}\mathbf{r}^{(n)} \rangle} \mathbf{r}^{(n)}, \quad \mathbf{r}^{(n)} = \mathbf{f} - \mathbf{A}\mathbf{u}^{(n)}, \quad n \geq 0. \quad (5.2.13)$$

There, the user is released from an appropriate choice of the damping parameter, at the price of a slightly more expensive iteration step. The exact steepest descent iteration has the same error reduction factor $q = \frac{\kappa(\mathbf{A})-1}{\kappa(\mathbf{A})+1}$ *as the Richardson iteration with optimal damping parameter* α^*, *so that approximate versions of both schemes should exhibit comparable convergence results. Adaptive steepest descent iterations in the case of wavelet Riesz bases have been investigated in* [27, 64], *whereas the recent paper* [52] *studies the case of frames. In* [52], *an asymptotically optimal variant of* (5.2.13) *could be derived where the convergence and complexity estimates rely on the same assumptions as in Theorem 5.6.*

One may also think of an adaptive version of the conjugate gradient iteration for symmetric positive definite systems. However, first numerical experiments using wavelet frame discretizations indicate that such a method does not really pay compared to the simpler steepest descent scheme. A similar observation has also been made in [64] *for the case of wavelet bases.*

5.3   Numerical Experiments

For the verification of the convergence and complexity of the **SOLVE** algorithm, we choose some test examples where adaptive schemes pay off most, i.e., where the exact solution has a significantly higher regularity in the Besov scale rather than in the Sobolev scale. Essentially, the following one– and two–dimensional benchmark examples are identical to those considered in [52].

5.3.1   1D Experiments

As a test example in one space dimension, we consider the variational formulation of the Poisson equation on the unit interval $\Omega = (0, 1)$

$$-u'' = f \text{ in } \Omega, \quad u(0) = u(1) = 0. \quad (5.3.1)$$

The operator $A = -\frac{\mathrm{d}^2}{\mathrm{d}x^2}$ has order $2t = 2$. In order to obtain an exact solution u that has limited Sobolev regularity, we choose as a right–hand side the functional $f(v) := 4v(\frac{1}{2}) + \int_0^1 g(x)v(x)\,\mathrm{d}x$, where

$$g(x) = -9\pi^2 \sin(3\pi x) - 4.$$

Then u is given by the superposition

$$u(x) = -\sin(3\pi x) + \begin{cases} 2x^2, & x \in [0, \tfrac{1}{2}) \\ 2(1-x)^2 & x \in [\tfrac{1}{2}, 1] \end{cases},$$

see Figure 5.2.

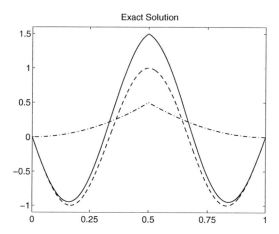

Exact Solution

Figure 5.2: Exact solution u (solid line) for the one–dimensional test example being the sum of the dashed and dash–dotted functions.

On the one hand, u is contained in $H^s(\Omega) \cap H^1_0(\Omega)$ only for $s < \frac{3}{2}$. This means that linear methods can only converge with limited order. On the other hand, since u is continuous and piecewise smooth, it can be shown that $u \in B^s_\tau(L_\tau(\Omega))$ for *any* $s > 0$, $\tau = (s + \frac{1}{2})^{-1}$. By interpolation between $H^{1+\varepsilon}(\Omega)$ and $B^s_\tau(L_\tau(\Omega))$, we can derive that $u \in B^{2s+1}_\tau(L_\tau(\Omega))$ for any $s > 0$, $\tau = (s + \frac{1}{2})^{-1}$.

Consequently, the error of best N–term wavelet approximation of u in $H^1(\Omega)$ has a decay rate s that is only limited by the properties of the underlying wavelet system. By the above discussion, we can derive that for an aggregated spline wavelet Gelfand frame Ψ for $(H^1_0(\Omega), L_2(\Omega), H^{-1}(\Omega))$ with order of exactness m, there exists an expansion $u = F^* \mathbf{D}^{-1} \mathbf{u}$ with coefficient array $\mathbf{u} \in \ell^w_\tau$ for any $s < m-1$, $\tau = (s + \frac{1}{2})^{-1}$.

As an overlapping domain decomposition we choose $\Omega = \Omega_1 \cup \Omega_2$, where $\Omega_1 = (0, 0.7)$ and $\Omega_2 = (0.3, 1)$. Associated to this decomposition we construct a aggregated wavelet Gelfand frame according to Section 2.2, where we use all frame elements up to the scale $|\lambda| \leq j_{\max} = 12$. We consider the cases of piecewise linear ($m = 2$) and piecewise quadratic ($m = 3$) template wavelet bases $\Psi^\square$ from [126], with $\tilde{m} = m$ vanishing moments. Consequently, the optimal rates of convergence for adaptive frame schemes are bounded by $s < m - 1 \in \{1, 2\}$.

We have tested the algorithm **SOLVE** with the relaxation parameters $\alpha = 0.57$ for $m = \tilde{m} = 2$ and $\alpha = 0.4$ for $m = \tilde{m} = 3$. It should be noted that for these parameters α, the algorithm yielded the optimal convergence behavior, but we do

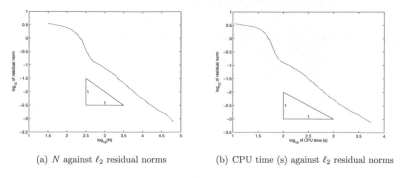

(a) N against ℓ_2 residual norms   (b) CPU time (s) against ℓ_2 residual norms

Figure 5.3: Convergence histories of **SOLVE** for the one–dimensional test example, using piecewise linear frame elements ($m = \tilde{m} = 2$).

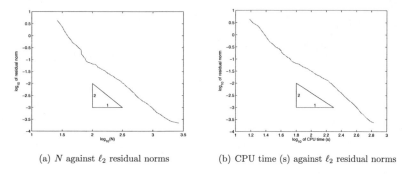

(a) N against ℓ_2 residual norms   (b) CPU time (s) against ℓ_2 residual norms

Figure 5.4: Convergence histories of **SOLVE** for the one–dimensional test example, using piecewise quadratic frame elements ($m = \tilde{m} = 3$).

not claim here that this choice of parameters matches the theoretically optimal relaxation parameter $\alpha^* = 2/(\lambda_{\max} + \lambda_{\min}^+)$. In fact, the latter constant may be difficult to estimate since the lowest eigenvalue $\lambda_{\min}^+$ of $\mathbf{A}_{|\,\mathrm{Ran}(\mathbf{A})}$ is hardly available in the case of frames. We refer to [142] for several case studies concerning the spectral behavior of $\mathbf{A}_{|\,\mathrm{Ran}(\mathbf{A})}$ and finite portions thereof in the case that the aggregated wavelet frame construction utilizes the wavelet bases from [61]. The remaining parameters in **SOLVE** are chosen to yield the quantitatively optimal response of the algorithm.

Generally speaking, the one–dimensional numerical experiments confirm the expected convergence rates and the asserted optimal complexity of **SOLVE**. In Figures 5.3 and 5.4, we plot the degrees of freedom $\#\operatorname{supp}\mathbf{u}^{(n)}$ and the CPU time against the ℓ_2 norm of the residuals $\mathbf{r}^{(n)}$. After an initialization phase, both the support sizes of the iterands and the associated computational work behave as predicted by the theory. Due to some caching strategies involved in the implementation, the graph of the CPU time attains the optimal slope not until some lower tolerance is reached.

Remark 5.10. *It should be noted that the concrete implementation of* **SOLVE** *used for the numerical experiments is not identical to Algorithm 5.3. This is due to the fact that the various parameters in* **SOLVE** *had been chosen to fulfill the worst case estimates of the error analysis of Theorem 5.4. An optimized parameter choice may then improve the quantitative response of the numerical scheme considerably.*

5.3.2 2D Experiments

In two spatial dimensions, we again consider the variational formulation of the Poisson equation

$$-\Delta u = f \text{ in } \Omega, \quad u_{|\partial\Omega} = 0. \tag{5.3.2}$$

The test problem is chosen in such a way that the application of adaptive algorithms pays off most, as it is the case for polygonal domains with reentrant corners. To this end, we choose Ω to be the L–shaped domain $\Omega = (-1,1)^2 \setminus [0,1)^2$. In a first test example, $u = S_{1,1}$ shall be the singularity function (3.0.7) corresponding to the reentrant corner at the origin. The solution u and the corresponding right–hand side $f = -\Delta u \in L_2(\Omega)$ are shown together in Figure 5.5.

It has already been stated in Chapter 3 that $u \in H^s(\Omega)$ only for $s < \frac{5}{3}$, but it is contained in every Besov space $B_\tau^{2s+1}(L_\tau(\Omega))$, where $s > 0$ and $\tau = (s + \frac{1}{2})^{-1}$, see Theorem 3.9. Consequently, the attainable convergence rate of a uniform refinement strategy is in general smaller than that of adaptive schemes.

As an overlapping domain decomposition we choose $\Omega = \Omega_1 \cup \Omega_2$, where $\Omega_1 = (-1,0) \times (-1,1)$ and $\Omega_2 = (-1,1) \times (-1,0)$, parametrized by the corresponding affine bijections $\kappa_i : \square \to \Omega_i$. Again Ψ shall be an aggregated wavelet Gelfand frame constructed as in Section 2.2, where we use all frame elements up to the scale $|\lambda| \leq j_{\max} = 7$. The underlying template wavelet basis $\Psi^\square$ on the unit square $\square$ is a tensor product of piecewise quadratic ($m = 3$) interval wavelet bases from [126] with $\tilde{m} = 3$ vanishing moments. As a consequence, the optimal rate of H^1–convergence for an adaptive frame scheme is bounded by $s < s^* = (3-1)/2 = 1$, whereas uniform refinement strategies would only converge with a rate of at most $(\frac{5}{3} - 1)/2 = \frac{1}{3}$.

We have tested the algorithm **SOLVE** with the relaxation parameter $\alpha = 0.24$, and again the remaining parameters are chosen to obtain the quantitatively optimal response of the algorithm. Figure 5.6 shows the asymptotic behavior of the degrees of freedom $\#\operatorname{supp}\mathbf{u}^{(n)}$ and the CPU time against the ℓ_2 norm of the residuals $\mathbf{r}^{(n)}$. Similar to the one–dimensional case, we observe that the numerical algorithm realizes the expected convergence rate $s = 1$ and the computational work behaves linearly in the number of unknowns as predicted by Theorem 5.6.

In the first test example, it has become visible that the adaptive frame algorithm does indeed recover singularities that are caused by the geometry of the underlying domain Ω. As a complement, we have also tested **SOLVE** in a situation where the singularity is induced by a nonsmooth right–hand side. Similar to the numerical experiments in [7], we fix a point x_0 from the interior of Ω and mimic the point eval-

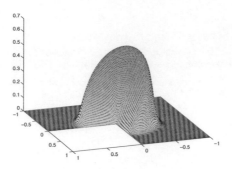

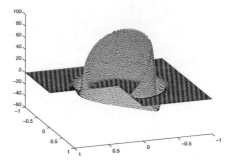

Figure 5.5: Exact solution (*top*) and right–hand side (*bottom*) for the first two–dimensional test example.

uation functional $v \mapsto v(x_0) \in H^{-1-\epsilon}(\Omega)$, $\epsilon > 0$, by setting the wavelet coefficients

$$\langle f, \psi_\lambda \rangle_{H^{-1}(\Omega) \times H_0^1(\Omega)} := \psi_\lambda(x_0), \quad \lambda \in \mathcal{J}. \tag{5.3.3}$$

We shall see in a moment that this coefficient array indeed gives rise to an element $f \in H^{-1}(\Omega)$. In the numerical example, we choose the point $x_0 := (-0.6, -0.6)$, and an approximation of the corresponding exact solution $u \in H_0^1(\Omega)$ on a fine scale is given in Figure 5.7.

First of all, the coefficients (5.3.3) are well–defined since the elements of the primal wavelet basis are globally continuous. By the multiresolution principles employed in the construction of Ψ, the point values $|\psi_\lambda(x_0)|$ scale like $2^{|\lambda|/2}$ as the level $|\lambda|$ tends to infinity. Hence, for any $\epsilon > 0$, we can infer from the uniform locality of the wavelet system that

$$\sum_{\lambda \in \mathcal{J}} 2^{-|\lambda|(1+\epsilon)} |\langle f, \psi_\lambda \rangle|^2 = \sum_{j \geq j_0} 2^{-j(1+\epsilon)} \sum_{\substack{\lambda : |\lambda| = j, \\ x_0 \in \mathrm{supp}(\psi_\lambda)}} |\psi_\lambda(x_0)|^2 \lesssim \sum_{j \geq j_0} 2^{-j(1+\epsilon)} 2^j < \infty.$$

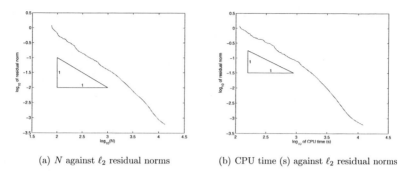

(a) N against ℓ_2 residual norms   (b) CPU time (s) against ℓ_2 residual norms

Figure 5.6: Convergence histories of **SOLVE** for the first two–dimensional test example, using piecewise quadratic frame elements $(m = \tilde{m} = 3)$.

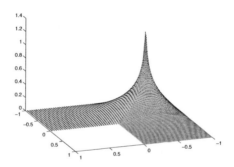

Figure 5.7: Exact solution for the second two–dimensional test example.

By an application of the Sobolev norm equivalence (1.2.16) and the density of Ψ, we see that f extends to an element of $H^{-1/2-\epsilon}(\Omega) \hookrightarrow H^{-1}(\Omega)$.

As regards the regularity properties of u, the conditions for Corollary 3.5 are not fulfilled since the right–hand side f is not contained in $L_2(\Omega)$. By Theorem 3.7, applied in the case $\mu = \frac{1}{2} - \epsilon$ and $\epsilon > 0$, we can only infer that $u \in B_\tau^\alpha(L_\tau(\Omega))$ for all $0 < \alpha < \frac{3}{2}$ and $\tau = (\frac{\alpha}{2} + \frac{1}{p})^{-1}$. Despite the fact that this is not the scale of best N–term approximation in H^1, we may still hope to observe a distinct convergence rate in a numerical experiment. In fact, the residual errors then drop at a convergence rate of 1, at least for moderate tolerances, see Figure 5.8. For low target accuracies, the convergence gets more indetermined. Due to the restriction to all frame elements below a certain level $j_{\max}$, an additional spatial error has been introduced which affects the numerical results much earlier than in the first test example. For low target accuracies, the measured residuals do no longer correspond to the true ones. Nevertheless, we again see that the computational work scales linearly in the number of degrees of freedom.

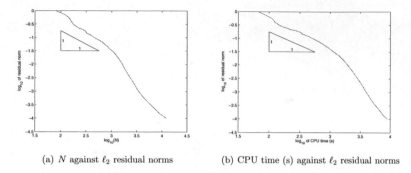

(a) N against ℓ_2 residual norms   (b) CPU time (s) against ℓ_2 residual norms

Figure 5.8: Convergence histories of **SOLVE** for the second two–dimensional test example, using piecewise quadratic frame elements ($m = \tilde{m} = 3$).

Part III

Discretization of Parabolic Problems

Chapter 6

Linear Parabolic Problems

This chapter is focused on the theoretical background of the parabolic problems under consideration, in particular concerning existence, uniqueness and regularity properties of the corresponding solution.

In Section 6.1, we review the theory of analytic semigroups. It turns out that a solution v of the homogeneous problem (0.0.20) always exists, given as the orbit of the initial value u_0 under the action of an analytic semigroup $\{e^{tA}\}_{t\geq 0}$. The temporal regularity of v hinges on the regularity of u_0, measured in the scale of fractional domains $D(A^\alpha)$ of A. Section 6.2 will then be centered around the solution of the inhomogeneous problem (0.0.21). We collect the basic properties of the inhomogeneous solution part w which is given as a convolution of the forcing term f with the semigroup $\{e^{tA}\}_{t\geq 0}$. Since (0.0.21) does not always hold in the sense of strong differentiability, we have to introduce the concept of a *mild* solution. Finally, in Section 6.3 we discuss the spatial reguarity properties of $u(t)$.

6.1   Analytic Semigroups

In this section, we are concerned with the construction of a solution v of the homogeneous initial value problem (0.0.20). We are especially interested in the case where the differential equation $v'(t) = Av(t)$ holds in the sense of strong differentiability. It turns out that under a sectoriality condition on A, such a v can be defined using the theory of analytic semigroups. For a more detailed survey of the topic we refer to the textbooks [2, 115, 125, 138].

6.1.1   Properties of Sectorial Operators

Let us start with the definition of sectorial operators. Given a Banach space X, a linear, densely defined operator $A : D(A) \subset X \to X$ is called *sectorial*, if there are constants $z_0 \in \mathbb{R}$, $\omega_0 \in (\frac{\pi}{2}, \pi)$ and $M > 0$, such that the resolvent set $\rho(A)$ contains the open sector

$$\Sigma_{z_0,\omega_0} := \big\{ z \in \mathbb{C} \setminus \{z_0\} : |\arg(z - z_0)| < \omega_0 \big\}, \tag{6.1.1}$$

and the resolvent operator $R(\lambda, A) := (\lambda I - A)^{-1}$ of A is bounded in norm by

$$\left\| R(z, A) \right\|_{\mathcal{L}(X)} \leq \frac{M}{|z - z_0|}, \quad z \in \Sigma_{z_0, \omega_0}. \tag{6.1.2}$$

Note that in contrast to other definitions of sectoriality that can be found in the literature, we explicitly added the density of $D(A)$ since we will work with densely defined operators A only. In general, sectorial operators do not need to be densely defined, see [115] for details.

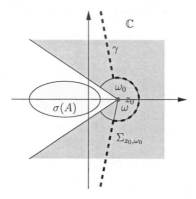

Figure 6.1: Spectral location for a sectorial operator A.

Since the resolvent set of a sectorial operator A is not empty, A is necessarily closed. As a consequence, the space $Y := D(A) \subset X$, equipped with the *graph norm*

$$\|x\|_Y := \|x\|_X + \|Ax\|_X, \quad x \in Y, \tag{6.1.3}$$

is a Banach space.

As is visualized in Figure 6.1, the spectrum $\sigma(A)$ of a sectorial operator A is also contained in a sector

$$\Sigma = \mathbb{C} \setminus \Sigma_{z_0, \omega_0} = \left\{ z \in \mathbb{C} : |\arg(z - z_0)| \geq \omega_0 \right\} \cup \{z_0\}.$$

Hence, by the holomorphic functional calculus, it is possible to define the operator $f(A)$ for a large class of holomorphic functions defined on Σ. Two prominent examples for the function f shall be discussed in the following. Firstly, the entire function $f(z) = e^{tz}$ leads to the analytic semigroup e^{tA}. Secondly, the function $f(z) = (-z)^{-\alpha}$ for $\alpha > 0$, being holomorphic on $\mathbb{C} \setminus \mathbb{R}_+$, will be used to define the fractional powers $(-A)^{-\alpha}$ of operators A with spectrum $\sigma(A)$ in the left half plane of $\mathbb{C}$.

For $t > 0$, the operator exponential $e^{tA} : X \to X$ of a sectorial operator A is given by means of the Dunford integral

$$e^{tA} := \frac{1}{2\pi i} \int_\gamma e^{tz} R(z, A) \, dz, \quad t > 0. \tag{6.1.4}$$

Here, see again Figure 6.1, γ can be any curve of the form

$$\gamma = \big\{ z \in \mathbb{C} : |\arg(z-z_0)| = \omega, |z-z_0| \geq r \big\} \cup \big\{ z \in \mathbb{C} : |\arg(z-z_0)| \leq \omega, |z-z_0| = r \big\} \tag{6.1.5}$$

for $r > 0$ and $\omega \in (\frac{\pi}{2}, \omega_0)$, oriented counterclockwise. For the basic definition and properties of the Banach space–valued integral (6.1.4), we refer the reader to the textbooks [74, 138].

Note that the definition of e^{tA} is independent of the concrete choice of r and ω, since the mapping $z \mapsto \mathrm{e}^{tz} R(z, A) \in \mathcal{L}(X)$ is holomorphic in the sector Σ_{z_0, ω_0}. By the definition of the Dunford integral, the integral e^{tA} converges in the uniform operator topology over X, and hence $\mathrm{e}^{tA} \in \mathcal{L}(X)$. As a complement, we set

$$\mathrm{e}^{0A} x := x, \quad x \in X, \tag{6.1.6}$$

i.e., $\mathrm{e}^0 = \mathrm{id}_X$. Then, for a sectorial operator $A : D(A) \to X$, the family $\{\mathrm{e}^{tA}\}_{t \geq 0} \subset \mathcal{L}(X)$ is called the *analytic semigroup generated by A in X*.

Generally speaking, a family of linear operators $\{T(t)\}_{t \geq 0} \subset \mathcal{L}(X)$ is said to be a *semigroup*, if $T(0) = I$ and $T(t+s) = T(t)T(s)$ for $s, t \geq 0$. The semigroup law of $\{\mathrm{e}^{tA}\}_{t \geq 0}$ as well as some other basic properties are collected in the following theorem, see [115, Proposition 2.1.1]:

Theorem 6.1. *Let $\{\mathrm{e}^{tA}\}_{t \geq 0}$ be the analytic semigroup generated by the sectorial operator $A : D(A) \to X$. Then the following assertions are valid:*

(i) *For each $x \in X$, $k \in \mathbb{N}$ and $t > 0$, it is $\mathrm{e}^{tA} x \in D(A^k)$. If $x \in D(A^k)$, then $A^k \mathrm{e}^{tA} x = \mathrm{e}^{tA} A^k x$ holds for $t \geq 0$.*

(ii) *The semigroup law $\mathrm{e}^{tA} \mathrm{e}^{sA} = \mathrm{e}^{(t+s)A}$ holds, for all $s, t \geq 0$.*

(iii) *There are constants $M_k > 0$ for $k = 0, 1, \ldots$, such that*

$$\big\| t^k (A - z_0 I)^k \mathrm{e}^{tA} \big\|_{\mathcal{L}(X)} \leq M_k \mathrm{e}^{z_0 t}, \quad k \in \mathbb{N}_0, t > 0. \tag{6.1.7}$$

In particular, for any $\epsilon > 0$ and $k \in \mathbb{N}$, there is a constant $C_{k,\epsilon} > 0$, such that

$$\big\| t^k A^k \mathrm{e}^{tA} \big\|_{\mathcal{L}(X)} \leq C_{k,\epsilon} \mathrm{e}^{(z_0 + \epsilon)t}, \quad k \in \mathbb{N}, t > 0. \tag{6.1.8}$$

(iv) *The mapping $t \mapsto \mathrm{e}^{tA}$ belongs to $C^\infty\big((0, \infty); \mathcal{L}(X)\big)$ and it is $\frac{\mathrm{d}^k}{\mathrm{d}t^k} \mathrm{e}^{tA} = A^k \mathrm{e}^{tA}$ for all $t > 0$. Moreover, it has an analytic extension into the sector $\Sigma_{0, \omega_0 - \pi/2}$.*

If the spectrum $\sigma(A)$ is contained in the left half plane — as is the case for negative self–adjoint operators — the constant z_0 will be negative. Then, as a consequence of the estimate (6.1.7) for $k = 0$, we have exponential decay of the norms $\|\mathrm{e}^{tA} v\|_X$ for $v \in X$ and $t \to \infty$. The effect of (6.1.7) for arbitrary $k \in \mathbb{N}_0$ is also known as *parabolic smoothing*.

As becomes visible in part (iv) of Theorem 6.1, the mapping

$$v : [0, T] \to X, \quad v(t) := \mathrm{e}^{tA} u_0 \tag{6.1.9}$$

fulfills the differential equation $v'(t) = Av(t)$ in the sense of strong differentiability for all $t > 0$. Moreover, we have of course that $v(0) = u_0 \in X$. Functions with these properties shall be given a special name to in the sequel. Namely, for a given forcing term $f \in C([0, T]; X)$ and $u_0 \in X$, we call a function $u : [0, T] \to X$ a *classical solution* of (0.0.19), if $u \in C^1((0, T]; X) \cap C((0, T]; D(A)) \cap C([0, T]; X)$, $u(0) = u_0$, and $u'(t) = Au(t) + f(t)$ holds for all $t \in (0, T]$.

The additional smoothness conditions for v from (6.1.9) can be verified by an application of the following theorem about the behavior of an analytic semigroup at $t = 0$, see [115, Proposition 2.1.4]:

Theorem 6.2. *Let $\{e^{tA}\}_{t \geq 0}$ be the analytic semigroup generated by the sectorial operator $A : D(A) \to X$. Then the following assertions are valid:*

(i) For all $x \in X$, it is $\lim_{t \to 0+} e^{tA} x = x$.

(ii) For all $x \in X$ and $t \geq 0$, the integral $\int_0^t e^{sA} x \, ds$ belongs to $D(A)$ and

$$A \int_0^t e^{sA} x \, ds = e^{tA} x - x. \tag{6.1.10}$$

If, additionally, the function $s \mapsto Ae^{sA} x$ belongs to $L_1([0, t]; X)$, then it is

$$e^{tA} - x = \int_0^t Ae^{sA} x \, ds. \tag{6.1.11}$$

(iii) For all $x \in D(A)$, it is $\lim_{t \to 0+}(e^{tA} x - x)/t = Ax$.

(iv) For all $x \in D(A)$, it is $\lim_{t \to 0+} Ae^{tA} x = Ax$.

Corollary 6.3. *For all $u_0 \in X$, the mapping v from (6.1.9) is the classical solution of the homogeneous problem (0.0.20).*

Proof. The existence of $v' = Av(t)$ for all $t > 0$ has been verified in Theorem 6.1. Moreover, the continuity of $v' : (0, T] \to X$ follows from part (iv) of Theorem 6.2. For a fixed $t > 0$, we have $w := e^{tA} u_0 \in D(A)$ by part (i) of Theorem 6.1. Given $s \geq t$, it is then

$$\|v(s) - v(t)\|_{D(A)} = \|Ae^{(s-t)A} w - Aw\|_X + \|e^{(s-t)A} w - w\|_X$$

which converges to zero as $s \to t$ by parts (i) and (iii) of Theorem 6.2. Hence v is continuous at t as a mapping into $D(A)$ and as a mapping into X. The continuity of $v : [0, T] \to X$ at $t = 0$ follows from part (i) of Theorem 6.2. $\square$

A slightly stronger solution concept covers situations where u is also differentiable at $t = 0$. More precisely, for a given forcing term $f \in C([0, T]; X)$ and $u_0 \in X$, we call a function $u : [0, T] \to X$ a *strict solution* of (0.0.19), if $u \in C^1([0, T]; X) \cap C([0, T]; D(A))$, $u(0) = u_0$, and $u'(t) = Au(t) + f(t)$ holds for all $t \in [0, T]$.

Then one can immediately draw the following conclusion:

Corollary 6.4. *For all $u_0 \in D(A)$, the mapping v from (6.1.9) is the strict solution of the homogeneous problem (0.0.20).*

Proof. By part (iii) of Theorem 6.2, we know that $v : [0, T] \to X$ is also differentiable at $t = 0$ with

$$v'(0) = \lim_{t \to 0+} (v(t) - v(0))/t = Au_0 = Av(0),$$

and part (iv) of Theorem 6.2 guarantees that $v'(t) = Av(t) \in X$ is continuous at $t = 0$. Analogously, we can infer the continuity of $v : [0, T] \to D(A)$ at $t = 0$. □

It becomes clear that the global temporal regularity of v is completely determined by the smoothness of the initial value u_0. This observation is also imminent when considering the initial values to be in the domain of higher powers of A:

Corollary 6.5. *Assume that $u_0 \in D(A^k)$ for $k \in \mathbb{N}$. Then the mapping v from (6.1.9) is contained in $C^k([0, T]; X) \cap C^{k-1}([0, T]; D(A))$.*

Proof. We proceed by induction over k. The case $k = 1$ has been considered in Corollary 6.4. Assume then that the claim holds for k and let $u_0 \in D(A^{k+1})$. Setting $w_0 := Au_0 \in D(A^k)$, we know by induction hypothesis that $w(t) := e^{tA}w_0$ is the strict solution of

$$w'(t) = Aw(t), \quad w(0) = w_0,$$

with $w \in C^k([0, T]; X) \cap C^{k-1}([0, T]; D(A))$. Since

$$v'(t) = Ae^{tA}u_0 = e^{tA}Au_0 = e^{tA}w_0 = w(t)$$

holds for all $t \geq 0$ due to part (i) of Theorem 6.1, the claim follows. □

We will discuss this general principle in more detail in Subsection 6.1.3.

6.1.2 Examples: Sectorial Operators Given by a Form

In order to translate the abstract definition of sectorial operators and analytic semigroups into a more concrete setting, one may consider the special case of sectorial operators $A : D(A) \subset X \to X$ that are realizations of elliptic operators, as discussed in Chapter 0. This approach is also known as the *form method*, see [4, 105].

To this end, let $H \hookrightarrow X$ be continuously and densely embedded Hilbert spaces. Moreover, assume that $a : H \times H \to \mathbb{C}$ is a sesquilinear form that is continuous in the sense of (0.0.6) and *coercive*, i.e., the *Gårding-type inequality*

$$\operatorname{Re} a(v, v) + C_1 \|v\|_X^2 \geq C_2 \|v\|_H^2, \quad v \in H, \tag{6.1.12}$$

holds with constants $C_1 \in \mathbb{R}$, $C_2 > 0$. Note that these assumptions are indeed fulfilled in the situation sketched in Chapter 0. Then one can associate a linear, potentially unbounded operator $A : D(A) \subset H \to X$ to a by the setting

$$D(A) := \{v \in H : \exists w =: Av \in X \text{ such that } a(v, \varphi) = \langle w, \varphi \rangle_X, \ \varphi \in H\}. \tag{6.1.13}$$

Since H is dense in X, the operator A is well-defined on $D(A)$. If X happens to coincide with H', then it is $D(A) = H$ and the definition of A in (6.1.13) coincides with (0.0.8), up to the sign. In most cases, however, X will be chosen to be an intermediate pivot space between H and H'. This is preferably done in such a way that X is identified with its normed dual X' so that $H \hookrightarrow X \hookrightarrow H'$ yields a Gelfand triple.

Under the aforementioned assumptions on the sesquilinear form a, it can be shown that the negative operator $-A : D(A) \subset X \to X$ is indeed sectorial, cf. [138, Section 2]:

Lemma 6.6. Let $a : H \times H \to \mathbb{C}$ be continuous and coercive. Moreover, let V be a Hilbert space in which H is continuously and densely embedded and consider the Gelfand triple $H \hookrightarrow V \hookrightarrow H'$. Then, for both choices $X \in \{V, H'\}$, the corresponding operator $-A : D(A) \subset X \to X$ is sectorial, with parameters $z_0 = C_1$ and some $\omega_0 \in (\frac{\pi}{2}, \pi]$.

Example 6.7. For $X \in \{H^{-1}(\Omega), L_2(\Omega)\}$ and $H = H_0^1(\Omega)$, we can define the symmetric bilinear form

$$a(v, w) = \int_\Omega \nabla v(x) \nabla w(x) \, \mathrm{d}x, \quad v, w \in H. \tag{6.1.14}$$

It is known that on H, a is bounded and coercive with

$$a(v, v) \geq C_2 \|v\|_{H^1(\Omega)}^2, \quad v \in H_0^1(\Omega) \tag{6.1.15}$$

for a constant $C_2 > 0$. As already mentioned in Chapter 0, the sectorial operator associated with a is called the Dirichlet Laplacian $-A = \Delta_\Omega^{\mathrm{D}}$. In the case $X = H^{-1}(\Omega)$, it is $D(A) = H_0^1(\Omega)$. For $X = L_2(\Omega)$, as laid out in Chapter 3, one can show that $D(A) = H^2(\Omega) \cap H_0^1(\Omega)$ only if Ω is convex or has a sufficiently smooth boundary $\partial\Omega$. For arbitrary Lipschitz or polygonal domains Ω, $D(A)$ is strictly larger and no immediate identification with Sobolev or Besov spaces is known. The domains of higher powers of A are discussed in Section 6.3.

It is also possible to consider the bilinear form (6.1.14) on the full space $H^1(\Omega)$ or other closed subspaces thereof, leading to the *Neumann Laplacian* operators. However, in this thesis we shall restrict the discussion to the case of full Dirichlet boundary conditions.

6.1.3   Fractional Powers of Sectorial Operators

As we have seen in Subsection 6.1.1, the temporal regularity of $v(t) = \mathrm{e}^{tA} u_0$ at $t = 0$ strongly depends on the regularity of the initial value u_0, measured in the scale of spaces $D(A^k) \hookrightarrow X$, $k \in \mathbb{N}$. It is possible to refine this observation by the introduction of intermediate spaces between $D(A^k)$ and X.

One reasonable choice are of course the real interpolation spaces $[X, D(A^k)]_{\theta, q}$, see [11] or Section 1.2. These spaces would fit well into the setting of elliptic operator equations where one considers Sobolev and Besov spaces that are also interpolation spaces. However, it turns out that interpolation spaces are not always fully compabible with the mapping properties of the analytic semigroup. More natural intermediate spaces are given by the domains of *fractional* powers of $-A$ that shall be introduced in the sequel.

As already stated above, for a sectorial operator A with a spectrum $\sigma(A)$ contained in the open left half plane $\{z \in \mathbb{C} : \operatorname{Re} z < 0\}$, we have $0 \in \rho(A)$ and hence $z_0 < 0$ (6.1.1). Then we can employ the holomorphic functional calculus to define the negative fractional powers

$$(-A)^{-\alpha} := \frac{1}{2\pi i} \int_\gamma (-z)^{-\alpha} R(z, A) \, \mathrm{d}z, \quad \alpha > 0. \tag{6.1.16}$$

Here the integration path γ is chosen as in (6.1.5), avoiding the positive real axis. The convergence of (6.1.16) in $\mathcal{L}(X)$ is guaranteed by the resolvent bound estimate (6.1.2). An equivalent representation of $(-A)^{-\alpha}$ is

$$(-A)^{-\alpha} = \frac{1}{\Gamma(\alpha)} \int_0^\infty t^{\alpha-1} \mathrm{e}^{tA} \, \mathrm{d}t, \tag{6.1.17}$$

with the Euler Γ function

$$\Gamma(x) = \int_0^\infty \mathrm{e}^{-xt} t^{x-1} \, \mathrm{d}t, \quad x > 0, \tag{6.1.18}$$

see [115]. Note that for $\alpha = 1$, the definition of $(-A)^{-1}$ from (6.1.16) coincides with the inverse $R(0, A)$, since the resolvent of any sectorial operator A has the integral representation [115, Lemma 2.1.6]

$$R(z, A) - \int_0^\infty \mathrm{e}^{-tz} \mathrm{e}^{tA} \, \mathrm{d}t, \quad \operatorname{Re} z > z_0. \tag{6.1.19}$$

Moreover, due to the fact that $(-A)^{-\alpha}(-A)^{-\beta} = (-A)^{-(\alpha+\beta)}$, the operator $(-A)^{-\alpha}$ is a bijection onto its range $\operatorname{Ran}((-A)^{-\alpha}$, see [115, Lemma 2.2.13]. As a consequence, the inverse $(-A)^\alpha := \left((-A)^{-\alpha}\right)^{-1}$ is well-defined as an unbounded operator from $D\left((-A)^\alpha\right) := \operatorname{Ran}\left((-A)^{-\alpha}\right) \subset X$ to X, and we set $(-A)^0 := \operatorname{id}_X$. For $\alpha > 0$, the operator $(-A)^\alpha$ is closed, and we can equip $D\left((-A)^\alpha\right)$ with the norm $\|(-A)^\alpha \cdot \|_X$ under which it is a Banach space. If $\alpha \geq \beta > 0$, then $D\left((-A)^\alpha\right) \hookrightarrow D\left((-A)^\beta\right)$. For integer values of $\alpha = k$, the spaces $D\left((-A)^\alpha\right)$ coincide with the domains $D(A^k)$ of the corresponding power of A.

Theorem 6.8 ([125, Th. 2.6.13]). *Let A be the infinitesimal generator of an analytic semigroup e^{tA}, such that $0 \in \rho(A)$. Then:*

(i) $\mathrm{e}^{tA} : X \to D\left((-A)^\alpha\right)$ for all $t > 0$ and $\alpha \geq 0$.

(ii) For every $\alpha \in \mathbb{R}$ and $x \in D\left((-A)^\alpha\right)$, we have $\mathrm{e}^{tA}(-A)^\alpha x = (-A)^\alpha \mathrm{e}^{tA} x$.

(iii) For all $t > 0$ and $\alpha \in \mathbb{R}$, the operator $(-A)^\alpha \mathrm{e}^{tA}$ is bounded with

$$\|(-A)^\alpha \mathrm{e}^{tA}\|_{\mathcal{L}(X)} \leq M_\alpha t^{-\alpha} \mathrm{e}^{\beta t}, \tag{6.1.20}$$

where $\beta > \sup\{\operatorname{Re} z : z \in \sigma(A)\}$.

(iv) If $0 < \alpha \leq 1$ and $x \in D\left((-A)^\alpha\right)$, then it is

$$\|\mathrm{e}^{tA} x - x\|_X \leq C_\alpha t^\alpha \|(-A)^\alpha x\|_X. \tag{6.1.21}$$

As a consequence of (6.1.21), we can expect that the solution v of the homogeneous problem is at least in $C^\alpha([0,T];X) \cap C^{\alpha-1}([0,T];D(A))$ whenever $u_0 \in D((-A)^\alpha)$. Further details on the behavior of e^{tA} in interpolation spaces can be found in [115, Section 2.2].

For $0 < \alpha < 1$, the interpolation spaces $[X, D(A)]_{\theta,q}$ and the domains of fractional powers $(-A)^\alpha$ are related. In [115, 139], it is shown that

$$\big[X, D(A)\big]_{\alpha,1} \hookrightarrow D\big((-A)^\alpha\big) \hookrightarrow \big[X, D(A)\big]_{\alpha,\infty}. \tag{6.1.22}$$

These embeddings are not sharp in general. As a complement to (6.1.22), for some special cases including self–adjoint negative definite operators A on a Hilbert space, one has the equivalence

$$D\big((-A)^\alpha\big) \simeq \big[X, D(A)\big]_{\alpha,2}. \tag{6.1.23}$$

6.2   Mild Solutions of Inhomogeneous Problems

In this section we will review the construction of a solution to the inhomogeneous problem (0.0.21). It is well–known that the existence of classical or strict solutions with strong temporal differentiability is not guaranteed for an arbitrary forcing term f. Under the integrability assumption $f \in L_1([0,T];X)$, we can at least define a solution candidate by the variation of constants formula

$$u(t) = e^{tA}u_0 + \int_0^t e^{(t-s)A} f(s) \, ds, \quad t \geq 0. \tag{6.2.1}$$

The function u from (6.2.1) is called the *mild solution* of (0.0.19). This definition is motivated by the fact that for $f \in L_1([0,T];X) \cap C\big((0,T];X\big)$ and $u_0 \in X$, the classical solution u of (0.0.19) also matches formula (6.2.1), see [115, Prop. 4.1.2] for a proof. Hence classical solutions are also mild. Moreover, it is known that mild solutions solve (0.0.19) in an integral sense:

Proposition 6.9 ([115, Prop. 4.1.5]). *Assume that $f \in L_1([0,T];X)$, $u_0 \in X$ and let u be the mild solution* (6.2.1) *to* (0.0.19)*. Then for every $t \in [0,T]$, it is $\int_0^t u(s)\,ds \in D(A)$ and we have the identity*

$$u(t) - u_0 = A \int_0^t u(s) \, ds + \int_0^t f(s) \, ds, \quad t \in [0,T]. \tag{6.2.2}$$

Unfortunately, for arbitrary integrable forcing terms f, the mild solution (6.2.1) might not be a classical or even strict one. Necessary and sufficient conditions for a mild solution to be classical or strict are collected in the following lemma, see [115] for a proof:

Lemma 6.10. *Let $f \in L_1([0,T];X) \cap C\big((0,T];X\big)$, $u_0 \in X$ and u be the mild solution* (6.2.1)*. Then the following conditions are equivalent:*

(i) u is a classical solution of (0.0.19)*.*

(ii) $u \in C\big((0,T]; D(A)\big)$

(iii) $u \in C^1\big((0,T]; X\big)$

If in addition $f \in C\big([0,T]; X\big)$, *then the following conditions are equivalent:*

(i) u *is a strict solution of* (0.0.19).

(ii) $u \in C\big([0,T]; D(A)\big)$

(iii) $u \in C^1\big([0,T]; X\big)$

In particular, if the right–hand side f is only in $L_1([0,T]; X)$, we do not know whether u maps into the space $D(A)$ for $t > 0$. However, under the slightly stronger condition $f \in L_2([0,T]; X)$, this is indeed the case, see [103, 112]:

Theorem 6.11. *Let A be sectorial with $z_0 < 0$, $u_0 = 0$ and $f \in L_2([0,T]; X)$. Then we have $u \in L_2\big([0,T]; D(A)\big) \cap H^1([0,T]; X)$, where u is the mild solution u from* (6.2.1).

The additional condition on the square integrability of f that ensures the spatial regularity $u(t) \in D(A)$ is not crucial, since we will have to raise further regularity assumptions on f anyway to guarantee also temporal smoothness of u.

Since the homogeneous solution part has already been discussed in Section 6.1, we can restrict the discussion of temporal regularity to the mild solution

$$w(t) = \int_0^t e^{(t-s)A} f(s)\, ds, \quad t \geq 0 \tag{6.2.3}$$

of the inhomogeneous problem (0.0.21). It is immediate by the convolution structure of (6.2.3) that any temporal smoothness assumption on f like boundedness or Hölder regularity should carry over to analogous properties of w. Let us mention at least two well–known results in this direction:

Theorem 6.12 ([115, Prop. 4.2.1]). *Let $f \in L_\infty([0,T]; X)$. Then, for every $\alpha \in (0,1)$, it is $w \in C^\alpha([0,T]; X) \cap C([0,T]; [X, D(A)]_{\alpha,1})$. More precisely, it is $w \in C^{1-\alpha}([0,T]; [X, D(A)]_{\alpha,1})$ and the estimate*

$$\|w\|_{C^{1-\alpha}([0,T];[X,D(A)]_{\alpha,1})} \leq C \|f\|_{L_\infty([0,T];X)} \tag{6.2.4}$$

holds with C independent from f.

Theorem 6.13 ([125, Th. 3.5]). *Let $f \in C^\alpha([0,T]; X)$. Then the following statements hold true:*

(i) *For every $\delta > 0$, it is $Aw, w' \in C^\alpha([\delta,T]; X)$.*

(ii) *We have $Aw, w' \in C([0,T]; X)$.*

(iii) *If $f(0) = 0$, then $Aw, w' \in C^\alpha([0,T]; X)$.*

So, loosely speaking, we can expect that at least after an initial transient phase, the temporal smoothness of the solution u to the parabolic problem (0.0.19) is completely determined by that of the driving term f.

6.3 Spatial Regularity

It remains to collect the relevant spatial regularity properties of the solution u of (0.0.19). As we have already seen in the previous sections, the homogeneous solution part $v(t)$ is contained in the spaces $D(A^k)$, whereas a nontrivial forcing term f causes the inhomogeneous solution part w to map into $D(A)$. Hence in the sequel, we shall be engaged in the characterization of the spaces $D(A^k)$ in terms of Sobolev or Besov spaces, where A is a more concrete elliptic operator. To simplify the discussion, we will assume that $A = \Delta_\Omega^D$ is the Dirichlet Laplacian operator over some bounded domain Ω in $\mathbb{R}^d$, i.e., $-A$ is induced by the bilinear form a from (6.1.14).

In view of the numerical experiments, we will restrict the setting to the case of polygonal domains Ω in one or two spatial dimensions. Concerning the domain $D(A)$, we can hence utilize the regularity results for the Poisson equation that were recalled in Chapter 3. Generally speaking, the situation in two space dimensions is substantially different than in the one–dimensional case, as soon as the domain Ω has a nonsmooth boundary. Here the solution u exhibits a higher regularity in the scale of Besov spaces than in the classical Sobolev scale. Consequently, we shall treat the two situations separately.

6.3.1 The 1D Case

In the one–dimensional case $\Omega = (0,1)$, a full orthonormal system of eigenfunctions for the Dirichlet Laplacian $A = \Delta_\Omega^D = \frac{d}{dx^2}$ is known. To be precise, we have the L_2–normalized eigenfunctions

$$v_k(x) = \sqrt{2}\sin(k\pi x), \quad x \in (0,1), \ k \in \mathbb{N}_0 \tag{6.3.1}$$

and the corresponding eigenvalues

$$\lambda_k = -k^2\pi^2, \quad k \in \mathbb{N}_0. \tag{6.3.2}$$

By the Spectral Mapping Theorem for sectorial operators $A : D(A) \subset X \to X$ on a Banach space X, see, e.g., Corollary 2.3.7 of [115], we know that

$$\sigma(\mathrm{e}^{tA}) \setminus \{0\} = \mathrm{e}^{t\sigma(A)}, \quad t > 0. \tag{6.3.3}$$

As a consequence, since $A = \Delta_\Omega^D$ is negative selfadjoint with $\sigma(A) = \{\lambda_k\}_{k\in\mathbb{N}_0}$, we can derive the series expansion for any $\varphi \in L_2(0,1)$

$$\mathrm{e}^{tA}\varphi = \sum_{k\in\mathbb{N}} \mathrm{e}^{t\lambda_k}\langle\varphi, v_k\rangle_{L_2(0,1)}v_k, \quad t \geq 0. \tag{6.3.4}$$

In order to analyze the smoothing properties of e^{tA}, it is straightforward to explicitly estimate the higher derivatives

$$\left(\frac{d}{dx}\right)^\ell v_k(x) = \begin{cases} \sqrt{2}(-1)^m(k\pi)^\ell \sin(k\pi x) & , \ \ell = 2m \\ \sqrt{2}(-1)^m(k\pi)^\ell \cos(k\pi x) & , \ \ell = 2m+1 \end{cases},$$

which gives

$$\left\|\left(\frac{d}{dx}\right)^\ell v_k\right\|_{L_\infty(0,1)} \lesssim (k\pi)^\ell \tag{6.3.5}$$

and hence $\|e^{tA}\varphi\|_{W^\ell(L_\infty(0,1))} < \infty$ for any $\ell \in \mathbb{N}$ and all $t > 0$. Due to Theorem 6.1, we also know that $e^{tA}\varphi \in D(A) \hookrightarrow H_0^1(0,1)$ fulfills the Dirichlet boundary conditions for all $\varphi \in L_2(0,1)$ and $t > 0$. Alternatively, we can also iterate Theorem 3.1 to derive that $D(A^k) = H^{1+2k}(0,1) \cap H_0^1(0,1)$ which also implies $e^{tA}\varphi \in H^\ell(0,1)$ for all $\ell \in \mathbb{N}$ and all $t > 0$. Summing up, the solution $v(t) = e^{tA}u_0$ of the homogeneous problem (0.0.20) fulfills $v(t) \in H^s(0,1) \cap H_0^1(0,1)$ for any desired $s \geq 0$, even if the initial value u_0 is only in $L_2(0,1)$.

For a nontrivial forcing term f, the mild solution u of the initial value problem (0.0.19) will be given by the convolution integral (6.2.1). From the discussion of Section 6.2, we know that the inhomogeneous solution part w from (6.2.3) can be expected to map into $D(A)$, since f will in general not fulfill the boundary conditions. For the Dirichlet Laplacian operator, it is known that the space $D(A)$ coincides with $H^2(0,1) \cap H_0^1(0,1)$, since the underlying domain $\Omega = (0,1)$ is convex [104].

As a consequence, the superposition $u = v + w$ can be expected to have spatial H^2 regularity. In view of the numerical experiments, this means that for a numerical approximation of $u(t)$ for fixed t, algorithms based on a uniform space refinement should exhibit the same convergence rates as nonlinear approximation schemes, as long as $u(t)$ does not happen to have a Besov smoothness $B_\tau^{s+1}(L_\tau(0,1))$ with s strictly larger than 1 for other reasons, where $\tau = (s + \frac{1}{2})^{-1}$.

6.3.2 The 2D Case

For an arbitrary two–dimensional domain Ω, we cannot expect to have a full orthonormal eigensystem for the Dirichlet Laplacian at our disposal. Just to state an important example, the eigenfunctions and eigenvalues of the Dirichlet Laplacian on the L–shaped domain are only known numerically, see [12, 78] for details. Consequently, the spatial regularity analysis of the solution u will be substantially more delicate than in the univariate case.

As we have seen in Chapter 3, the characterization of $D(A)$ and more classical smoothness spaces like Sobolev or Besov spaces hinges on the regularity of the boundary $\partial\Omega$. This general fact also holds true when considering the domains of higher powers of A. If $\partial\Omega$ is C^∞, we may again simply iterate Theorem 3.1 to obtain $D(A^k) = H^{1+2k}(\Omega) \cap H_0^1(\Omega)$ and hence arbitrary high Sobolev smoothness of $u(t)$ for $t > 0$. For domains with a C^α boundary, $\alpha \geq 1$, we can still derive $D(A^k) \hookrightarrow H^s(\Omega)$ for $s = \min\{1 + 2k, \alpha\}$ from Theorem 3.1.

For the polygonal and non–convex domains Ω we are interested in, however, the L_2 Sobolev exponent of the functions in $D(A^k)$ will be limited by some value $s < 2$ by Theorems 3.3 and 3.8. In that case, we have already seen in Chapter 3 that the Besov regularity of functions from $D(A)$ is substantially higher, and we refer to the relevant Theorems 3.4 and 3.9 here. For the Besov regularity analysis of the domains of higher powers $D(A^k) = D(A^k; L_2(\Omega))$, to the author's knowledge there is no general result available in the literature. Only for the special case $k = 2$, one may iterate the above mentioned theorems twice to obtain, e.g., the following regularity result for the Dirichlet Laplacian on a Lipschitz domain.

Lemma 6.14. *Let $\Omega \subset \mathbb{R}^d$ be a bounded Lipschitz domain and let $A = \Delta_\Omega^D$ be the Dirichlet Laplacian over Ω. Then for all $1 < p \leq 2$, we have the continuous embedding*

$$D(A^2) \hookrightarrow B_\tau^\alpha(L_\tau(\Omega)), \quad 0 < \alpha < \min\left\{\frac{7}{2}, \frac{d}{d-1}\left(1 + \frac{1}{p}\right)\right\}, \quad \tau = \left(\frac{\alpha}{d} + \frac{1}{p}\right)^{-1},$$
(6.3.6)

where the embedding constant depends on α and p.

Proof. Given some $u \in D(A^2)$, it is $A^2 u = f$ for some $f \in L_2(\Omega)$. Hence, for $v := Au \in D(A)$, we know from Theorem 3.2 and the boundedness of Ω that $v \in H^{3/2}(\Omega) \hookrightarrow B_p^{3/2}(L_p(\Omega))$ for all $1 < p \leq 2$. Inserting v as a right–hand side into Theorem 3.4 with $\mu = \frac{3}{2} + 2 = \frac{7}{2} \geq 2 > 1 + \frac{1}{p}$, we get that $u \in B_\tau^\alpha(L_\tau(\Omega))$, $\tau = (\frac{\alpha}{d} + \frac{1}{p})^{-1}$, for the asserted range $0 < \alpha < \min\{\frac{d}{d-1}(1 + \frac{1}{p}), \frac{7}{2}\}$.   □

Remark 6.15. *In Lemma 1.27 of* [103] *it is claimed that* (6.3.6) *holds for the larger range $0 < \alpha < \min\{\frac{7}{2}, \frac{2d}{d-1}\}$. This is only possible if the metric parameter p is simultaneously chosen arbitrarily close to 1. Here we are particularly interested in the case that $d = 2$ and $p = 2$ is fixed, where we obtain the embedding $D(A^2) \hookrightarrow B_\tau^\alpha(L_\tau(\Omega))$ for $\alpha < 3$, $\tau = (\frac{\alpha}{2} + \frac{1}{2})^{-1}$.*

An analogous result to Lemma 6.14 can be derived for the special case of polygonal domains Ω in $\mathbb{R}^2$. Here one may use the expansion of $u \in D(A^2)$ into higher order singularity functions as in Theorem 3.10.

Lemma 6.16. *Let $\Omega \subset \mathbb{R}^2$ be a bounded polygonal domain and let $A = \Delta_\Omega^D$ be the Dirichlet Laplacian over Ω. Then we have the continuous embedding*

$$D(A^2) \hookrightarrow B_\tau^\alpha(L_\tau(\Omega)), \quad 0 < \alpha < \min\left\{3 + \frac{\pi}{\omega_j} : \omega_j > \pi\right\}, \quad \tau = \left(\frac{\alpha}{2} + \frac{1}{2}\right)^{-1}. \quad (6.3.7)$$

Proof. For $u \in D(A^2)$, it is $A^2 u = f$ for some $f \in L_2(\Omega)$. Hence, for $v := Au \in D(A)$, Theorem 3.8 gives that $v \in H^s(\Omega)$, $s < \check{s} := \min\{1 + \pi/\omega_j : \omega_j > \pi\}$. We insert v as a right–hand side into Theorem 3.10 and obtain that the regular part u_R of u is contained in $H^{\check{s}+2}(\Omega) \hookrightarrow B_{\check{\tau}}^{\check{s}+2}(L_{\check{\tau}}(\Omega))$ for all $\check{\tau} \leq 2$. The singular part u_S is contained in all Besov spaces $B_\tau^\alpha(L_\tau(\Omega))$, $\tau = (\frac{\alpha}{2} + \frac{1}{2})^{-1}$, $\alpha > 0$, see [44]. The claim follows from the superposition $u = u_R + u_S$.   □

Remark 6.17. *The argument in the proof of Lemma 6.16 is not sufficient to improve* (6.3.7) *in the case of $u \in D(A^k)$ for $k \geq 3$. This is due to the fact that $Au = v \in D(A^{k-1})$ is contained in $H^s(\Omega)$ only for $s < \min\{1 + \pi/\omega_j : \omega_j > \pi\}$.*

Based on Theorem 3.8, it is also possible to derive analogous decomposition results for the solution u of the heat equation over a bounded polygonal domain:

Theorem 6.18 ([85, Th. 5.2.1]). *Let $\Omega \subset \mathbb{R}^2$ be bounded, open and polygonal domain, and let $A = \Delta_\Omega^D$ be the Dirichlet Laplacian over Ω. Moreover, for $u_0 \in D(A)$ and $f \in L_2\big(\mathbb{R}_+; L_2(\Omega)\big)$, assume that u is the solution of (0.0.19). Then there exists $u_\mathrm{R} \in L_2\big(\mathbb{R}_+; H^2(\Omega)\big)$ with $u_\mathrm{R}' \in L_2\big(\mathbb{R}_+; L_2(\Omega)\big)$, and functions $\varphi_j \in H^{(1-\pi/\omega_j)/2}(\mathbb{R}_+)$, such that u can be decomposed into*

$$u(t) = u_\mathrm{R}(t) + \sum_{\omega_j > \pi} g_j(t) S_{j,1}. \tag{6.3.8}$$

Here $g_j(t) = g_j(t, r_j, \theta_j)$ is the convolution product on the positive halfline

$$g_j(t, r_j, \theta_j) := (E_j(\cdot, r_j, \theta_j) * \varphi_j)(t) = \int_0^t E_j(s, r_j, \theta_j) \varphi_j(t - s) \, \mathrm{d}s, \tag{6.3.9}$$

and $E_j(t) = E_j(t, r_j, \theta_j) = \frac{r_j}{2\sqrt{\pi t^3}} \mathrm{e}^{-r_j^2/(4t)}$ is given in polar coordinates (r_j, θ_j) with respect to the corner S_j.

Summing up the spatial regularity results in two space dimensions, we can expect the local exact solution $u(t)$ of the heat equation to have a significant higher regularity in the scale of Besov spaces $B_\tau^{2s+1}(L_\tau(\Omega))$, $\tau = (s + \frac{1}{2})^{-1}$, than in the Sobolev scale. Consequently adaptive strategies for the spatial approximation of $u(t)$ should pay off compared to algorithms based on a uniform space refinement, as soon as the underlying domain is non–convex and has a non–smooth boundary.

Chapter 7

Wavelet Discretization of Parabolic Problems

In this chapter, we turn to the development of an adaptive, wavelet–based numerical scheme for the linear parabolic equation (0.0.19). As already mentioned in the introduction, we shall follow Rothe's method which is also known as the horizontal method of lines. Doing so, the discretization is performed in two major steps. Firstly, we consider a semidiscretization in time, where we will employ an S–stage linearly implicit scheme. This approach is discussed in more detail in Section 7.1. We shall end up with an orbit of approximations $u^{(n)} \in L_2(\Omega)$ at intermediate times t_n that are implicitly given via the S elliptic stage equations. In a finite element context, this very approach has already been propagated in [107, 108].

For the realization of the increment $u^{(n)} \mapsto u^{(n+1)}$ and the spatial discretization of the stage equations, we will then employ the adaptive wavelet schemes introduced in Chapter 4 as a black box solver. In Section 7.2, we derive an adaptive increment algorithm and analyze its convergence and computational complexity properties.

7.1 Linearly Implicit Semidiscretization in Time

We shall now be concerned with the semidiscretization in time for the mild solution $u \in C([0,T];V)$ of the linear parabolic problem (0.0.19). In order to obtain a convenient notation and for potential generalizations of the discussed scheme towards nonlinear problems, we will consider (0.0.19) in the generalized form

$$u'(t) = F(t, u(t)), \quad t \in (0, T], \quad u(0) = u_0, \tag{7.1.1}$$

where $F : [0, T] \times H \to H'$ is given as

$$F(t, v) = Av + f(t), \quad t \in [0, T], \quad v \in H. \tag{7.1.2}$$

Hence we have $\partial_v F(t, v) = A$ and $\partial_t F(t, v) = f'(t)$. In view of the specific spatial discretization in Section 7.2, we assume that the sectorial operator A under consideration is induced by a symmetric bilinear form.

117

7.1.1 ROW–Methods

As already stated in the introduction of this chapter, we consider an S–stage linearly implicit method for the semidiscretization in time. By this we mean an iteration of the form

$$u^{(n+1)} = u^{(n)} + h \sum_{i=1}^{S} b_i k_i \qquad (7.1.3)$$

with the *stage equations*

$$(I - h\gamma_{i,i}J)k_i = F\left(t_n + \alpha_i h, u^{(n)} + h \sum_{j=1}^{i-1} \alpha_{i,j} k_j\right) + hJ \sum_{j=1}^{i-1} \gamma_{i,j} k_j + h\gamma_i g, \quad i = 1, \ldots, S,$$

$$(7.1.4)$$

where we set

$$\alpha_i := \sum_{j=1}^{i-1} \alpha_{i,j}, \quad \gamma_i := \sum_{j=1}^{i} \gamma_{i,j}. \qquad (7.1.5)$$

The operator $I - h\gamma_{i,i}J$ in (7.1.4) has to be understood as a boundedly invertible operator from H to H', with the equality (7.1.4) in the sense of H'. Such a scheme is also known as a method of *Rosenbrock* type, see [94, 137] for details. All the quantities h, J, k_i and g in (7.1.4) do of course depend on the time step number n, but we drop the index n here for readability. The coefficients b_i, $\alpha_{i,j}$ and $\gamma_{i,j}$ have to be suitably chosen according to the desired properties of the Rosenbrock method. We will turn to this question in the next subsection.

As a special case of (7.1.4), a *Rosenbrock–Wanner method* or *ROW–method* results if one chooses the exact derivatives $J = \partial_v F(t_n, u^{(n)})$ and $g = \partial_t F(t_n, u^{(n)})$. In this thesis, we will confine the setting to these ROW–type methods.

Remark 7.1. *The specific choice of J and g in a ROW–method is not needed to derive a convergent time discretization, cf. [94]. In the larger class of W–methods, J is allowed to be an mere approximation to the exact Jacobian $\partial_v F(t_n, u^{(n)})$. Moreover, one often chooses $\gamma_i = 0$ there, so that g does not even enter the algorithm. This is done at the cost of additional order conditions and a more complicated stability analysis. Using W–methods, the system matrix in a discretization of (7.1.4) can be thinned out further, which may also be of interest in a wavelet setting. The particular analysis of wavelet–based W–methods is still in its infancy [5] and goes beyond the scope of this thesis.*

Example 7.2. *The most simple Rosenbrock method is the* linearly implicit Euler method, *where $u^{(n+1)} = u^{(n)} + hk_1$ and $(I - hJ)k_1 = F(t_n, u^{(n)})$, see [21, 94]. For the schemes used in the numerical experiments, we refer to Table 8.1.*

In the sequel, we will only consider Rosenbrock schemes with coincident diagonal entries $\gamma_{i,i} = \gamma > 0$. This is not a critical limitation since almost all popular Rosenbrock schemes in the literature have this property. The matching of the diagonal entries $\gamma_{i,i}$ can be exploited in practical realizations of the increment (7.1.3), see Section 7.2, since then also the operators $I - h\gamma_{i,i}J$ of the S stage equations coincide.

By definition, the practical realization of a Rosenbrock method only requires the successive solution of S linear operator equations per time step. Fortunately, iterative Newton methods as needed in Runge–Kutta methods are not necessary here since the Jacobian of the right–hand side F is worked into the integration formula of a Rosenbrock scheme. In practice, a Rosenbrock scheme will be implemented in a slightly different way than given by (7.1.4). Introducing the variable $u_i :=
h \sum_{j=1}^{i} \gamma_{i,j} k_j$, the additional application of the operator J in the right–hand side of (7.1.4) can be avoided by rewriting (7.1.4) as

$$\left(\frac{1}{h\gamma_{i,i}} I - J \right) u_i = F \left(t_n + \alpha_i h, u^{(n)} + \sum_{j=1}^{i-1} a_{i,j} u_j \right) + \sum_{j=1}^{i-1} \frac{c_{i,j}}{h} u_j + h\gamma_i g, \quad i = 1, \ldots, S,$$

(7.1.6)

and

$$u^{(n+1)} = u^{(n)} + \sum_{i=1}^{S} m_i u_i \tag{7.1.7}$$

where we have used the coefficients

$$\Gamma = (\gamma_{i,j})_{i,j=1}^{S}, \tag{7.1.8}$$

$$(a_{i,j})_{i,j=1}^{S} = (\alpha_{i,j})_{i,j=1}^{S} \Gamma^{-1}, \tag{7.1.9}$$

$$(c_{i,j})_{i,j=1}^{S} = \mathrm{diag}(\gamma_{1,1}^{-1}, \ldots, \gamma_{S,S}^{-1}) - \Gamma^{-1}, \tag{7.1.10}$$

$$(m_1, \ldots, m_S)^{\top} = (b_1, \ldots, b_S)^{\top} \Gamma^{-1}. \tag{7.1.11}$$

For the special case of a *linear* parabolic problem (0.0.19), the stage equations to be solved can then be recast as follows:

$$\left(\frac{1}{h\gamma_{i,i}} I - J \right) u_i = A \left(u^{(n)} + \sum_{j=1}^{i-1} a_{i,j} u_j \right) + f(t_n + \alpha_i h) + \sum_{j=1}^{i-1} \frac{c_{i,j}}{h} u_j + h\gamma_i g, \quad i = 1, \ldots, S,$$

(7.1.12)

where $J \approx A$ and $g \approx f'(t_n)$ as above.

7.1.2   Convergence Results

When applied to an initial value problem (7.1.1) for *ordinary* differential equations, i.e., for a finite–dimensional space H, the Rosenbrock method realizes a (classical) convergence order $p \in \mathbb{N}$ if the global error behaves like

$$e_n := u^{(n)} - u(t_n) = \mathcal{O}(h^p), \quad h \to 0 \tag{7.1.13}$$

uniformly over $[0, T]$, with $h_n = t_{n+1} - t_n \leq h$. Estimates of this type can be expected to hold for sufficiently regular right–hand sides F whenever the coefficients of the Rosenbrock scheme satisfy the corresponding algebraic order conditions [94, 137].

For the convergence analysis of the scheme (7.1.4) applied to *partial* differential equations, one needs further stability properties of the Rosenbrock method under consideration. In order to obtain sufficient convergence criteria, one considers the associated *stability function*

$$R(z) = 1 + z\mathbf{b}^{\top}(\mathbf{I} - z\mathbf{B})^{\top}\mathbf{1}, \tag{7.1.14}$$

of the Rosenbrock method, where $\mathbf{b} = (b_i)_{i=1}^{S}$, $\mathbf{B} = (\beta_{i,j})_{1 \leq j \leq i \leq S}$, $\beta_{i,j} = \alpha_{i,j} + \gamma_{i,j}$, $\alpha_{i,i} = 0$ and $\mathbf{1} = (1, \ldots, 1)^{\top}$. Essentially, R is the increment function in

$$u^{(1)} = R(\lambda h)u_0 \qquad (7.1.15)$$

when applying one step with stepsize h of the iteration (7.1.3) to the one–dimensional *Dahlquist model problem*

$$u'(t) = \lambda u(t), \quad u(0) = u_0. \qquad (7.1.16)$$

Consequently, $R(z)$ is a rational approximation to the exponential function e^z. A convergence order p in (7.1.13) corresponds to the relationship $e^z - R(z) = \mathcal{O}(z^{p+1})$ as z tends to zero.

Then, a Rosenbrock method with stability function R is called $A(\theta)$–*stable*, if $|R(z)| \leq 1$ for all $z \in \mathbb{C}$ with $|\arg(z)| \geq \pi - \theta$. If, additionally, the limit value $|R(\infty)|$ is strictly smaller than 1, the method is called *strongly* $A(\theta)$–*stable*. Stiff components in the solution will be damped out rapidly if the method is L–*stable*, i.e., $R(\infty) = 0$.

For applications in partial differential equations, it turns out that the classical integer convergence order p of a given Rosenbrock method can no longer be achieved in general. In the case of low classical orders $p \leq 2$, convergence results for ordinary differential equations indeed carry over also to the infinite–dimensional case. Under the assumption that the stage equations are solved exactly, the following result concerning the temporal convergence of Rosenbrock methods was proved in [114]:

Theorem 7.3. *For $\theta > \pi - \omega_0$, where ω_0 is given by the sectoriality condition (6.1.1), consider a strongly $A(\theta)$–stable Rosenbrock method of order $p \geq 2$. Suppose that (7.1.1) has a unique solution u with temporal derivatives $u'' \in L_2([0,T]; H)$ and $u''' \in L_2([0,T]; H')$. Then, for sufficiently small time step sizes $h \leq h_0$, there exists a unique numerical solution u_n, $0 \leq nh \leq T$, with the error bound*

$$\left(h \sum_{n=0}^{N} \|e_n\|_H^2 \right)^{1/2} + \max_{0 \leq n \leq N} \|e_n\|_V \lesssim h^2 \left(\int_0^T \|u''(t)\|_H^2 \, dt + \int_0^T \|u'''(t)\|_{H'}^2 \, dt \right)^{1/2}.$$

$$(7.1.17)$$

In contrast to Theorem 7.3, for arbitrary Rosenbrock methods with a higher classical order $p \geq 3$ one is in general faced with a phenomenon called *order reduction*. Here, in the application of the scheme to an initial value problem for partial differential equations, one might observe a fractional order $p' \in [0, p]$ of convergence. The attainable value of p' is influenced by the spatial regularity of the solution u, measured in the scale of spaces $D(A^\alpha)$. This is due to the fact that higher powers of the unbounded operator A are present in the local truncation error. Order reduction effects were observed first in the application of Runge–Kutta methods to partial differential equations in [96, 120, 128]. For the application of Rosenbrock methods to linear parabolic problems, we refer the reader to [121]. In Theorem 2 from loc. cit., sufficient conditions were derived under which a higher order of convergence $p' \geq 3$ may be expected:

Theorem 7.4. *For $\theta > \pi - \omega_0$ and a given $A(\theta)$–stable Rosenbrock method of order p with stability function R and $|R(\infty)| \neq 1$, define the rational functions*

$$W_2(z) := \frac{\mathbf{b}^\top (I - z\mathbf{B})^{-1}(\alpha^2 - 2\mathbf{B}^2 \mathbf{1})}{1 - R(z)}, \qquad (7.1.18)$$

$$W_k(z) := \frac{\mathbf{b}^\top (I - z\mathbf{B})^{-1}(\alpha^k - k\mathbf{B}\alpha^{k-1})}{1 - R(z)}, \quad k \geq 3, \qquad (7.1.19)$$

and let $q \in \mathbb{N}$ be the maximal integer such that

$$W_k(z) = 0, \quad 2 \leq k \leq q, \quad \text{for all } z \text{ with } |\arg(z)| \geq \pi - \omega, \qquad (7.1.20)$$

or $q = 1$ if (7.1.20) is empty. Here we set $\alpha^k := (\alpha_i^k)_{i=1,\dots,s}$, $\omega \in (\theta, \pi - \omega_0)$ is fixed and ω_0 is given by the sectoriality condition (6.1.1). Then, for a given linear parabolic problem (0.0.19), the Rosenbrock method is stable and for a discretization with constant stepsizes, the global error behaves like

$$\|u(nh) - u_n\|_{L_2} = \mathcal{O}(h^{p'}), \quad p' = \min\{p, q + 2 + \bar{\nu}\}, \qquad (7.1.21)$$

where $\bar{\nu}$ is given by

$$\bar{\nu} := \sup\{\nu \in \mathbb{R} : f^{(j)}(t) \in D(A^\nu) \text{ for } t \in [0, T], \ 0 \leq j \leq p\}. \qquad (7.1.22)$$

It has already been observed in [120] that for the heat equation $u_t = u_{xx} + g(t)$ on the unit interval with homogeneous Dirichlet boundary conditions, the order of L_2–convergence of an $A(\theta)$–stable Rosenbrock is bounded from below by

$$p' \geq \min\{p, 3.25\}. \qquad (7.1.23)$$

More generally, since it is always $q \geq 1$ in Theorem 7.4, we should observe a numerical order of convergence of at least $p' \geq \min\{p, 3\}$ also in the two–dimensional case. Hence for the numerical examples discussed in Chapter 8, order reduction phenomena will only affect fourth order schemes.

Remark 7.5. *In the case of nonlinear problems, the problem of order reduction for Rosenbrock methods is more severe. For nonlinear parabolic equations with homogeneous Dirichlet boundary conditions, the numerical order of convergence of an arbitrary strongly $A(\theta)$–stable Rosenbrock scheme in general drops down to a value $p' = 2 + \beta$ with $\beta = \frac{3}{4} - \epsilon$ and arbitrarily small $\epsilon > 0$. For Neumann or inhomogeneous boundary conditions, p' may attain even smaller values, see [114] for details. Fortunately, it is possible to construct special Rosenbrock methods that fulfill the conditions (7.1.20) at least for $q = 2$ and lead to schemes of full convergence order 3 independent of the spatial regularity, see [3, 109] for some recent results.*

7.1.3 Stepsize Control

As already mentioned, we are interested in the adaptive solution of the parabolic problem (0.0.19), i.e., we are looking for approximations $u^{(n)} \approx u(t_n)$, such that the discretization error $\|u^{(n)} - u(t_n)\|_V$ at intermediate points $t_n \in [0, T]$ stays below some prescribed tolerance ε. By the nature of the analytic problem (0.0.19), see also Chapter 6, a constant temporal stepsize h is of course not the most economic choice to do so. At least for times t close to 0 and in situations where the driving term f is not smooth at t, it is advisable to choose small values of h in order to track the behavior of the exact solution correctly. In regions where f and u are temporally smooth, larger time step sizes may be used. Generally speaking, the current value of h should be as small as possible to ensure the desired accuracy but also sufficiently large to avoid unnecessary computational cost. As a consequence, we have to employ an a posteriori temporal error estimator to control the current value of h. Moreover, since the iteration (7.1.3) takes place in a Hilbert space, it cannot be implemented exactly and we have to take additional spatial errors $\tilde{u}^{(n+1)} - u^{(n+1)}$ into account. However, under the assumption that the spatial discretization is also done adaptively, we can interpret the spatial perturbation as a controllable additional error of the temporal discretization so that the step size selection will be nearly independent from the actual spatial discretization.

Ideally, an adaptive algorithm for non–stationary problems should be based on an a posteriori estimator for the *global* discretization error. However, the control of global errors is a difficult problem that has not been solved in full generality so far, even in the case of ordinary initial value problems. In recent years, the estimation of the global discretization error via adjoint problems has gained popularity, see [110], yet the practical advantages over the classical approach of *local* error estimation still have to be verified. The traditional approach does not try to control the global temporal discretization error but resorts to estimators for the local truncation error at t_n

$$\delta_h(t_n) := \Phi^{t_n, t_n + h}(u(t_n)) - u(t_n + h), \qquad (7.1.24)$$

where $\Phi^{t_n, t_n + h} : V \to V$ is the increment mapping of the given Rosenbrock scheme at time t_n with stepsize h. For the global error at $t = t_{n+1} = t_n + h_n$, we have the decomposition

$$e_{n+1} = u^{(n+1)} - u(t_{n+1}) = \Phi^{t_n, t_n + h_n}(u^{(n)}) - \Phi^{t_n, t_n + h_n}(u(t_n)) + \delta_{h_n}(t_n), \qquad (7.1.25)$$

i.e., e_{n+1} comprises the local error at time t_n and the difference between the current Rosenbrock step $\Phi^{t_n, t_n + h_n}(u^{(n)})$ and the virtual step $\Phi^{t_n, t_n + h_n}(u(t_n))$ with starting point $u(t_n)$. For stable one–step methods, the latter term can be understood as a propagated error term from the previous step. Of course, the propagation of local errors is a potential problem for algorithms based on local error estimation. Especially in the case of badly conditioned initial value problems, a global error control is indispensable.

Taking a suitable norm $\|\cdot\|$ in (7.1.25), one obtains estimates of the form

$$\|e_{n+1}\| \leq \|e_n\| + \|\delta_{h_n}(t_n)\| \leq \|e_0\| + \sum_{k=0}^{n} \|\delta_{h_k}(t_k)\|. \qquad (7.1.26)$$

In practical computations, the norm $\|\cdot\|$ will typically be a combination of absolute and relative norms that involves also a scaling of the solution components, see [93, 107]. An accepted choice of $\|\cdot\|$ is to use a weighted root mean square norm. The corresponding discrete variant in the wavelet coefficient space reads as

$$\|\mathbf{d}_k\| := \left(\frac{1}{\# \operatorname{supp} \mathbf{d}_k} \sum_{\lambda \in \operatorname{supp} \mathbf{d}_k} \frac{|d_{\lambda,k}|^2}{|w_\lambda|^2} \right)^{1/2}, \qquad (7.1.27)$$

where the wavelet coefficient array $\mathbf{d}_k$ belongs to the current L_2 error estimate $\delta_k = \mathbf{d}_k^\top \Psi \approx \delta_{h_k}(t_k)$. Here we use the weights $w_\lambda := \texttt{ATOL} + \texttt{RTOL} \cdot \max\{|c_\lambda^{(n)}|, |c_\lambda^{(n+1)}|\}$, and $u^{(n)} = (\mathbf{c}^{(n)})^\top \Psi$ with $\mathbf{c}^{(n)} = (c_\lambda^{(n)})_{\lambda \in \mathcal{J}}$. The parameters $\texttt{ATOL}$ and $\texttt{RTOL}$ are chosen according to the desired target accuracy ε.

Estimators for the local discretization error $\delta_{h_n}(t_n)$ can be either based on an embedded lower order scheme or on extrapolation techniques, see [93, 94]. It has turned out that a local error estimation based on extrapolation techniques is slightly more expensive but exhibits a better performance for very low tolerances. For applications to partial differential equations, only moderate accuracy requirements are present. In this case, embedding strategies yield comparable results and thus are our method of choice. In principle, the error estimator is then given by a suitable norm of the difference $u^{(n+1)} - \hat{u}^{(n+1)}$, where $\hat{u}^{(n+1)}$ is an alternative increment of order less or equal to $p - 1$ corresponding to a different coefficient set $\hat{b} = (\hat{b}_i)_{i=1}^S$ in (7.1.3). The computation of $\hat{u}^{(n+1)}$ causes no additional cost since the same stage solutions k_i are considered as for $u^{(n+1)}$. Concerning the concrete step size algorithm, we follow [107] and use an improved version of the standard step size controller from [93] as propagated by Gustafsson et al. [90, 91].

7.2 Spatial Discretization with Wavelet Methods

Since the iteration (7.1.3) cannot be implemented numerically, we will address the numerical approximation of all the ingredients by finite–dimensional counterparts in this section. Precisely, we have to find approximate, computable iterands $\tilde{u}^{(n+1)}$, such that the additional error $\tilde{u}^{(n+1)} - u^{(n+1)}$ introduced by the spatial discretization stays below some given tolerance ε when measured in an appropriate norm. Hence this perturbation of the virtual orbit $\{u^{(n)}\}_{n \geq 0}$ can be interpreted as a controllable additional error of the temporal discretization. The accumulation of local perturbations in the course of the iteration is then an issue for the stepsize controller. In order not to spoil the convergence behavior of the unperturbed iterands $u^{(n)}$ in (7.1.17), we will demand that $\tilde{u}^{(n+1)} - u^{(n+1)}$ stays small in the topology of H, which results in the requirement

$$\|\tilde{u}^{(n+1)} - u^{(n+1)}\|_H \leq \varepsilon \qquad (7.2.1)$$

for the numerical scheme, where $\varepsilon > 0$ is the desired target accuracy. Since the exact iterands $u^{(n)}$ in (7.1.3) are contained in $H \hookrightarrow V$, we shall utilize an appropriate wavelet basis for these spaces. Precisely, we will make the following assumption:

(W) We assume that $\Psi = \{\psi_\lambda\}_{\lambda \in \mathcal{J}}$ is a wavelet Riesz basis in V with dual basis $\tilde{\Psi} = \{\tilde{\psi}_\lambda\}_{\lambda \in \mathcal{J}}$, so that the rescaled version $\mathbf{D}_0^{-1}\Psi$ is also a Riesz basis in H, where $(\mathbf{D}_0)_{\lambda,\lambda} := |\langle A\psi_\lambda, \psi_\lambda \rangle|^{1/2}$ is the energy norm of ψ_λ.

As discussed in Chapter 1, it is indeed possible to construct such a wavelet basis on a bounded polygonal domain $\Omega \subset \mathbb{R}^d$. Unless otherwise stated, all sequence spaces ℓ_p and ℓ_τ^w in this section refer to the overall wavelet index set $\mathcal{J}$.

7.2.1 Properties of the Exact Increment

For the following analysis, let $\mathbf{u}^{(n)} := \langle u^{(n)}, \tilde{\Psi} \rangle^\top \in \ell_{2,\mathbf{D}_0}$ be the primal wavelet coefficient array of the exact iterand $u^{(n)} = (\mathbf{D}_0 \mathbf{u}^{(n)})^\top \mathbf{D}_0^{-1} \Psi \in H \hookrightarrow V$. Before we discuss how to ensure the error estimate (7.2.1), let us first analyze the algebraic operations needed to compute the exact increment coefficients $\mathbf{u}^{(n+1)}$.

Observe that by (7.1.7), the exact increment $u^{(n+1)}$ differs from $u^{(n)}$ by a linear combination of the exact solutions u_i of the S stage equations (7.1.12). The operators involved in (7.1.12) take the form

$$B_\alpha := \alpha I - A, \quad \alpha \geq 0, \tag{7.2.2}$$

where $\alpha = (h\gamma_{i,i})^{-1}$ for the i–th stage equation. By the estimate

$$\langle B_0 v, v \rangle \leq \langle B_\alpha v, v \rangle = \alpha \langle v, v \rangle_V + \langle B_0 v, v \rangle \leq (C\alpha + 1)\langle B_0 v, v \rangle, \quad v \in H,$$

we see that the energy norms $\|v\|_{B_\alpha} := |\langle B_\alpha v, v \rangle|^{1/2}$ differ from $\|v\|_{B_0} \approx \|v\|_H$ only by an α–dependent constant:

$$\|v\|_{B_0} \leq \|v\|_{B_\alpha} \leq (C\alpha + 1)^{1/2} \|v\|_{B_0}, \quad v \in H. \tag{7.2.3}$$

As a consequence, we can state the following lemma:

Lemma 7.6. *Let Assumption* (W) *hold and define*

$$(\mathbf{D}_\alpha)_{\lambda,\lambda} := \|\psi_\lambda\|_{B_\alpha}, \quad \lambda \in \mathcal{J}. \tag{7.2.4}$$

Then the system $\mathbf{D}_\alpha^{-1}\Psi$ *is a Riesz basis in the energy space* $(H, \|\cdot\|_{B_\alpha})$, *with Riesz constants independent from* $\alpha \geq 0$:

$$\|\mathbf{c}\|_{\ell_2} \approx \|\mathbf{c}^\top \mathbf{D}_\alpha^{-1}\Psi\|_{B_\alpha}, \quad \mathbf{c} \in \ell_2. \tag{7.2.5}$$

Proof. Using Assumption (W), we can compute

$$\begin{aligned}
\|\mathbf{c}^\top \mathbf{D}_\alpha^{-1}\Psi\|_{B_\alpha}^2 &= \alpha\|\mathbf{c}^\top \mathbf{D}_\alpha^{-1}\Psi\|_V^2 + \|\mathbf{c}^\top \mathbf{D}_\alpha^{-1}\Psi\|_{B_0}^2 \\
&\approx \alpha\|\mathbf{D}_\alpha^{-1}\mathbf{c}\|_{\ell_2}^2 + \|\mathbf{D}_\alpha^{-1}\mathbf{D}_0\mathbf{c}\|_{\ell_2}^2 \\
&= \sum_{\lambda \in \mathcal{J}} \frac{\alpha + \|\psi_\lambda\|_{B_0}^2}{\|\psi_\lambda\|_{B_\alpha}^2} |c_\lambda|^2
\end{aligned}$$

for all sequences $\mathbf{c} \in \ell_2$, with constants independent from α. Since $\|\psi_\lambda\|_V \approx 1$ by Assumption (W), the claim (7.2.5) immediately follows from the estimate

$$\frac{\alpha + \|\psi_\lambda\|_{B_0}^2}{\|\psi_\lambda\|_{B_\alpha}^2} \approx \frac{\alpha\|\psi_\lambda\|_V^2 + \|\psi_\lambda\|_{B_0}^2}{\|\psi_\lambda\|_{B_\alpha}^2} = 1.$$

$\square$

Remark 7.7. *It is also possible to use an α-dependent diagonal scaling matrix $\tilde{\mathbf{D}}_\alpha$ of the form $(\tilde{\mathbf{D}}_\alpha)_{\lambda,\lambda} := \alpha \mid 2^{|\lambda|t}$, which again results in a Riesz basis $\tilde{\mathbf{D}}_\alpha^{-1}\Psi$ for $(H, \| \cdot \|_{B_\alpha})$. However, in practical computations a diagonal scaling involving the energy norm generally yields tighter Riesz bounds than a scaling with $\tilde{\mathbf{D}}_\alpha$.*

Lemma 7.6 ensures that we can use the Riesz basis $\mathbf{D}_\alpha^{-1}\Psi$, $\alpha = (h\gamma_{i,i})^{-1}$ as test functions in a variational formulation of (7.1.12). Abbreviating the exact right–hand side of (7.1.12) by

$$r_{i,h} := A\Big(u^{(n)} + \sum_{j=1}^{i-1} a_{i,j}u_j\Big) + f(t_n + \alpha_i h) + \sum_{j=1}^{i-1} \frac{c_{i,j}}{h}u_j + h\gamma_i f'(t_n), \qquad (7.2.6)$$

we get the system of equations

$$\langle B_\alpha u_i, \mathbf{D}_\alpha^{-1}\Psi \rangle^\top = \langle r_{i,h}, \mathbf{D}_\alpha^{-1}\Psi \rangle^\top. \qquad (7.2.7)$$

Inserting a wavelet representation of $u_i = (\mathbf{D}_\alpha u_i)^\top \mathbf{D}_\alpha^{-1}\Psi$ into the variational formulation (7.2.7), we end up with the biinfinite linear system in ℓ_2

$$\mathbf{D}_\alpha^{-1}\langle B_\alpha \Psi, \Psi \rangle^\top \mathbf{D}_\alpha^{-1} \mathbf{D}_\alpha u_i = \mathbf{D}_\alpha^{-1}\langle r_{i,h}, \Psi \rangle^\top. \qquad (7.2.8)$$

As a consequence of the uniform Riesz basis property (7.2.5), the spectral condition numbers of the diagonally preconditioned system matrices

$$\mathbf{B}_\alpha := \mathbf{D}_\alpha^{-1}\langle B_\alpha \Psi, \Psi \rangle^\top \mathbf{D}_\alpha^{-1} \qquad (7.2.9)$$

should stay uniformly bounded for varying values of $\alpha \geq 0$. Results of this type have already been addressed in [7, 33, 55].

For a quantitative study of $\kappa_2(\mathbf{B}_\alpha)$ in the case of the quadratic spline wavelet bases on the interval from [126], we refer the reader to Figure 7.1. For the range of parameters $0 \leq \alpha \leq 10^4$, which is sufficient for the numerical experiments, the spectral condition numbers $\kappa_2(\mathbf{B}_\alpha)$ stay below 20, outperforming the stabilized interval bases from [9].

In Figure 7.2, we plot the spectral condition numbers of $\mathbf{B}_\alpha$ in the case of a linear spline composite wavelet basis on the L–shaped domain Ω as constructed in [62], where we have applied the stabilization of the wavelets as in [9]. Also in the two–dimensional case, the diagonal preconditioning with $\mathbf{D}_\alpha$ suppresses the condition numbers below 70 for the desired range of α values. We note that the concrete values for $\kappa_2(\mathbf{B}_\alpha)$ we obtained for the composite basis on the L–shaped domain are comparable to the values reported in [9] for the case of interval bases. Moreover, they are significantly lower than the condition numbers of the L–domain bases constructed in [102].

The operators B_α being a linear combination of $-A$ and the identity operator, we may assume for the considered class of operators A that the biinfinite stiffness matrices $\mathbf{B}_\alpha$ are s^*–compressible with $s^* > \frac{m-t}{d}$, see [134] for details. Here we recall that, like in Chapter 1, m is the polynomial approximation order of the wavelet basis Ψ, $2t$ is the order of A and d the spatial dimension of Ω. The compressibility assumption indeed holds in the case that $A = \Delta_\Omega^\mathrm{D}$ is the Dirichlet Laplacian operator,

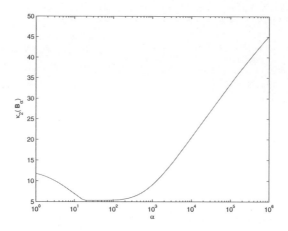

Figure 7.1: Spectral condition numbers $\kappa_2(\mathbf{B}_\alpha)$ for varying values of α and the wavelet bases on $\Omega = (0, 1)$ from [126], where $m = \tilde{m} = 3$.

i.e., for the treatment of the heat equation over the domain $\Omega \subset \mathbb{R}^d$, As a consequence of the s^*–compressibility, $\mathbf{B}_\alpha$ can be well approximated by sparse matrices, which is exploited in the next subsection.

In the sequel, we abbreviate with $\mathbf{G} := \langle \Psi, \Psi \rangle^\top$ the Gramian matrix of the wavelet basis Ψ in V. Then, by (7.2.6), the exact discrete right–hand side of the i–th stage equation

$$\mathbf{r}_{i,h} := \mathbf{D}_\alpha^{-1} \langle r_{i,h}, \Psi \rangle^\top, \quad \alpha = (h\gamma_{i,i})^{-1}, \tag{7.2.10}$$

decomposes into the sum

$$\mathbf{r}_{i,h} = - \mathbf{D}_\alpha^{-1} \mathbf{D}_0 \mathbf{B}_0 \Big(\mathbf{D}_0 \mathbf{u}^{(n)} + \mathbf{D}_0 \mathbf{D}_\alpha^{-1} \sum_{j=1}^{i-1} a_{i,j} \mathbf{D}_\alpha \mathbf{u}_j \Big) + \mathbf{D}_\alpha^{-1} \langle f(t_n + \alpha_i h), \Psi \rangle^\top$$

$$+ \mathbf{D}_\alpha^{-1} \mathbf{G} \mathbf{D}_\alpha^{-1} \sum_{j=1}^{i-1} \frac{c_{i,j}}{h} \mathbf{D}_\alpha \mathbf{u}_j + h\gamma_i \mathbf{D}_\alpha^{-1} \langle f'(t_n), \Psi \rangle^\top.$$

$$\tag{7.2.11}$$

Note that under the assumption $\gamma_{i,i} = \gamma$ for $i = 1, \ldots, S$, which holds true for each of the Rosenbrock schemes considered, we can use the same diagonal preconditioner $\mathbf{D}_\alpha^{-1}$ for each of the s stage solutions $\mathbf{u}_i$ belonging to the same time step. Hence no intermediate rescaling is needed.

7.2.2 An Approximate Increment Algorithm

Of course we cannot hope to know the exact values of $\mathbf{r}_{i,h}$ in practice, since the right–hand side involves evaluations of the driving terms and, in particular, biinfinite matrix–vector products for $i > 1$. Moreover, it will not be possible to solve

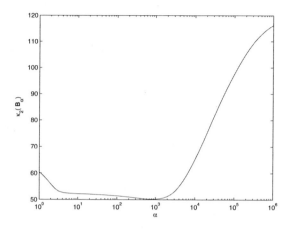

Figure 7.2: Spectral condition numbers $\kappa_2(\mathbf{B}_\alpha)$ for varying values of α and a composite wavelet basis on the L–shaped domain Ω, where $m = \tilde{m} = 2$.

the system (7.2.8) exactly. In this subsection, we therefore discuss the adaptive approximation $\tilde{\mathbf{u}}^{(n+1)}$ of the exact increment $\mathbf{u}^{(n+1)}$. Using the Riesz basis property (1.2.19), the requirement (7.2.1) is fulfilled if we can ensure that

$$\left\|\tilde{\mathbf{u}}^{(n+1)} - \mathbf{u}^{(n+1)}\right\|_{\ell_{2,\mathbf{D}_0}} \leq C_H^{-1}\varepsilon. \tag{7.2.12}$$

By the structure of the iteration (7.1.7), we essentially have to set up and to solve the S discrete stage equations (7.2.8) with specific tolerances depending on the target accuracy ε for the current time step. Since the right–hand side of the i–th stage equation involves the previous inexact stage solutions, we have to consider the effect of error propagation in a realization of the increment function.

Due to the fact that the involved operators $\mathbf{B}_\alpha$ are s^*–compressible, we can utilize the existence of adaptive numerical solvers for biinfinite systems like (7.2.8), e.g., those discussed in Chapter 4. Since $\mathbf{B}_\alpha$ is symmetric and positive definite, we argue against a Richardson–type iteration here and we opt for the more sophisticated CDD1 algorithm from [7, 33] instead. We abbreviate calls to the adaptive solver by the subroutine **SOLVE**, see also Theorem 4.12.

For the approximate evaluation of the driving term f we have to require the existence of a numerical routine

$$\mathbf{RHSF}[t, \varepsilon] \rightarrow \mathbf{f}_{t,\varepsilon}, \tag{7.2.13}$$

so that for $s < s^*$ it holds that

$$\left\|\mathbf{D}_0^{-1}\langle f(t), \Psi\rangle^\top - \mathbf{f}_{t,\varepsilon}\right\|_{\ell_2} \leq \varepsilon, \tag{7.2.14}$$

$$\# \operatorname{supp} \mathbf{f}_{t,\varepsilon} \lesssim |\mathbf{D}_0^{-1}\langle f(t), \Psi\rangle^\top|_{\ell_\tau^w}^{1/s}\varepsilon^{-1/s} \tag{7.2.15}$$

and the number or arithmetic operations to compute $\mathbf{f}_{t,\varepsilon}$ is bounded by a multiple of $\#\operatorname{supp}\mathbf{f}_{t,\varepsilon}$. Analogously, we assume to have a routine

$$\mathbf{RHSFP}[t,\varepsilon] \to \mathbf{f}'_{t,\varepsilon} \tag{7.2.16}$$

which approximates the coefficients $\mathbf{D}_0^{-1}\langle f'(t),\Psi\rangle^\top$ in ℓ_2 with

$$\left\|\mathbf{D}_0^{-1}\langle f'(t),\Psi\rangle^\top - \mathbf{f}'_{t,\varepsilon}\right\|_{\ell_2} \le \varepsilon, \tag{7.2.17}$$

$$\#\operatorname{supp}\mathbf{f}'_{t,\varepsilon} \lesssim |\mathbf{D}_0^{-1}\langle f'(t),\Psi\rangle^\top|_{\ell_\tau^w}^{1/s}\varepsilon^{-1/s} \tag{7.2.18}$$

and the computation of $\mathbf{f}'_{t,\varepsilon}$ takes only a constant times $\#\operatorname{supp}\mathbf{f}'_{t,\varepsilon}$ arithmetic operations

Having the three routines **SOLVE**, **RHSF** and **RHSFP** at hand, we are now in the position to specify the increment algorithm for the special case of ROW–methods, where $J = A$ and $g = f'(t_n)$ in (7.1.12):

Algorithm 7.8. ROW_INCREMENT$[\mathbf{D}_0\mathbf{u}^{(n)}, h, \varepsilon] \to \mathbf{D}_0\tilde{\mathbf{u}}^{(n+1)}$:
Let $\theta < 1/3$ be fixed.
$C := 2\|\mathbf{B}_\alpha^{-1}\|\max_{i,j}\left(\|\mathbf{B}_0\||a_{i,j}| + \|\mathbf{G}\|\frac{|c_{i,j}|}{h_{\min}}\right)$
$\eta_1 := \varepsilon(|m_1| + C\sum_{i=2}^S |m_i|\sum_{j=1}^{i-1}(1+C)^{i-1})^{-1}$
$\varepsilon_1 := \theta\eta_1/4$
$\varepsilon_{1,1} := \varepsilon_{1,2} := \theta\eta_1(4\|\mathbf{B}_\alpha^{-1}\|)^{-1}$
$\varepsilon_{1,4} := \theta\eta_1(4T|\gamma_1|\|\mathbf{B}_\alpha^{-1}\|)^{-1}$
$\mathbf{s}_{1,1} := \mathbf{APPLY}[\mathbf{B}_0, \mathbf{D}_0\mathbf{u}^{(n)}, \varepsilon_{1,1}]$
$\mathbf{s}_{1,2} := \mathbf{RHSF}[t_n + \alpha_1 h, \varepsilon_{1,2}]$
$\mathbf{s}_{1,4} := \mathbf{RHSFP}[t_n, \varepsilon_{1,4}]$
$\tilde{\mathbf{r}}_{1,h} := -\mathbf{D}_\alpha^{-1}\mathbf{D}_0\mathbf{s}_{1,1} + \mathbf{D}_\alpha^{-1}\mathbf{D}_0\mathbf{s}_{1,2} + h\gamma_1\mathbf{D}_\alpha^{-1}\mathbf{D}_0\mathbf{s}_{1,4}$
$\widetilde{\mathbf{D}_\alpha\mathbf{u}_1} := \mathbf{SOLVE}[\mathbf{B}_\alpha, \tilde{\mathbf{r}}_{1,h}, \varepsilon_1]$
For $i = 2,\ldots,S$ do
$\quad\varepsilon_i := \theta C(5\|\mathbf{B}_\alpha^{-1}\|)^{-1}\sum_{j=1}^{i-1}\eta_j$
$\quad\varepsilon_{i,1} := \varepsilon_{i,2} := \varepsilon_{i,3} := \theta C(5\|\mathbf{B}_\alpha^{-1}\|)^{-1}\sum_{j=1}^{i-1}\eta_j$
$\quad\varepsilon_{i,4} := \theta C(5T|\gamma_i|\|\mathbf{B}_\alpha^{-1}\|)^{-1}\sum_{j=1}^{i-1}\eta_j$
$\quad\mathbf{s}_{i,1} := \mathbf{APPLY}[\mathbf{B}_0, \mathbf{D}_0\mathbf{u}^{(n)} + \mathbf{D}_0\mathbf{D}_\alpha^{-1}\sum_{j=1}^{i-1}a_{i,j}\widetilde{\mathbf{D}_\alpha\mathbf{u}_j}, \varepsilon_{i,1}]$
$\quad\mathbf{s}_{i,2} := \mathbf{RHSF}[t_n + \alpha_i h, \varepsilon_{i,2}]$
$\quad\mathbf{s}_{i,3} := \mathbf{APPLY}[\mathbf{G}, \mathbf{D}_\alpha^{-1}\sum_{j=1}^{i-1}\frac{c_{i,j}}{h}\widetilde{\mathbf{D}_\alpha\mathbf{u}_j}, \varepsilon_{i,3}]$
$\quad\mathbf{s}_{i,4} := \mathbf{RHSFP}[t_n, \varepsilon_{i,4}]$
$\quad\tilde{\mathbf{r}}_{i,h} := -\mathbf{D}_\alpha^{-1}\mathbf{D}_0\mathbf{s}_{i,1} + \mathbf{D}_\alpha^{-1}\mathbf{D}_0\mathbf{s}_{i,2} + \mathbf{D}_\alpha^{-1}\mathbf{s}_{i,3} + h\gamma_i\mathbf{D}_\alpha^{-1}\mathbf{D}_0\mathbf{s}_{i,4}$
$\quad\widetilde{\mathbf{D}_\alpha\mathbf{u}_i} := \mathbf{SOLVE}[\mathbf{B}_\alpha, \tilde{\mathbf{r}}_{i,h}, \varepsilon_i]$
$\quad\eta_i := \varepsilon_i + \|\mathbf{B}_\alpha^{-1}\|\left(\varepsilon_{i,1} + \varepsilon_{i,2} + \varepsilon_{i,3} + T\gamma_i\varepsilon_{i,4} + \sum_{j=1}^{i-1}\left(\|\mathbf{B}_0\||a_{i,j}| + \|\mathbf{G}\|\frac{|c_{i,j}|}{h}\right)\eta_j\right)$
od
$\mathbf{D}_0\check{\mathbf{u}}^{(n+1)} := \mathbf{D}_0\mathbf{u}^{(n)} + \mathbf{D}_0\mathbf{D}_\alpha^{-1}\sum_{i=1}^S m_i\widetilde{\mathbf{D}_\alpha\mathbf{u}_i}, (1-\theta)\varepsilon$
$\mathbf{D}_0\tilde{\mathbf{u}}^{(n+1)} := \mathbf{COARSE}\left[\mathbf{D}_0\check{\mathbf{u}}^{(n+1)}, (1-\theta)\varepsilon\right]$

The goal of this subsection is then to analyze the convergence and complexity properties of the algorithm **ROW_INCREMENT**, as formulated in the following theorem:

Theorem 7.9. *Under the assumptions on the operator A stated above and for a given ROW–method, the routine* **ROW_INCREMENT** *outputs a finitely supported vector* $\mathbf{D}_0 \tilde{\mathbf{u}}^{(n+1)}$, *such that*

$$\|\mathbf{D}_0 \mathbf{u}^{(n+1)} - \mathbf{D}_0 \tilde{\mathbf{u}}^{(n+1)}\|_{\ell_2} \leq \varepsilon. \tag{7.2.19}$$

Moreover, we have

$$\# \operatorname{supp} \mathbf{D}_0 \tilde{\mathbf{u}}^{(n+1)} \lesssim |\mathbf{D}_0 \mathbf{u}^{(n+1)}|_{\ell_\tau^w}^{1/s} \varepsilon^{-1/s}, \tag{7.2.20}$$

and the number of arithmetic operations needed to compute $\mathbf{D}_0 \tilde{\mathbf{u}}^{(n+1)}$ *is bounded by a constant multiple of* $\tilde{C}^{1/s} \varepsilon^{-1/s}$, *where*

$$\tilde{C} := \max \left\{ |\mathbf{D}_0 \mathbf{u}^{(n)}|_{\ell_\tau^w}, |\mathbf{D}_0^{-1} \langle f'(t_n), \Psi \rangle^\top|_{\ell_\tau^w}, \max_{1 \leq i \leq S} |\mathbf{D}_0^{-1} \langle f(t_n + \alpha_i h), \Psi \rangle^\top|_{\ell_\tau^w} \right\}. \tag{7.2.21}$$

Before proving Theorem 7.9, let us first state a simple lemma about the boundedness of specific diagonal matrices like $\mathbf{D}_\alpha^{-1}$ and $\mathbf{D}_\alpha^{-1} \mathbf{D}_0$ on ℓ_τ and ℓ_τ^w spaces:

Lemma 7.10. *Let* $\mathbf{C} = \operatorname{diag}(c_\lambda)_{\lambda \in \mathcal{J}}$ *be a diagonal matrix, such that* $|c_\lambda| \leq 1$ *for all* $\lambda \in \mathcal{J}$. *Then* $\mathbf{C}$ *is bounded both on* ℓ_τ *for* $\tau > 0$ *and on* ℓ_τ^w *for* $0 < \tau < 2$, *with* $\|\mathbf{C}\|_{\mathcal{L}(\ell_\tau)} \leq 1$ *and* $\|\mathbf{C}\|_{\mathcal{L}(\ell_\tau^w)} \leq 1$.

Proof of Lemma 7.10. The boundedness of $\mathbf{C}$ on ℓ_τ with norm less or equal to 1 is trivial. In order to prove the boundedness of $\mathbf{C}$ on ℓ_τ^w, let $\gamma_n(\mathbf{v}) = v_{i_n}$ and $\gamma_n(\mathbf{Cv}) = c_{j_n} v_{j_n}$ be the n–th largest coefficients in modulus of $\mathbf{v} \in \ell_2$ and $\mathbf{Cv} \in \ell_2$, respectively. By definition, the indices i_n as well as the j_n are pairwise different. Then it is sufficient to show that

$$|c_{j_n} v_{j_n}| \leq |v_{i_n}|, \quad n \geq 1. \tag{7.2.22}$$

For $n = 1$, this is trivial since $|c_{j_1} v_{j_1}| \leq |v_{j_1}| \leq |v_{i_1}|$. Now assume that (7.2.22) holds for all $n < m$. If it were $|c_{j_m} v_{j_m}| > |v_{i_m}|$, then we could infer that

$$|v_{j_n}| \geq |c_{j_n} v_{j_n}| \geq |c_{j_m} v_{j_m}| > |v_{i_m}|, \quad n = 1, \ldots, m,$$

and hence that $j_1, \ldots, j_m \in \{i_1, \ldots, i_{m-1}\}$, which is obviously a contradiction to the indices j_n being pairwise distinct. $\square$

Proof of Theorem 7.9. For the proof of the estimate (7.2.19), recall the following basic fact about the solution of linear operator equations with a perturbed right-hand side. Namely, if $\mathbf{Bx} = \mathbf{y}$ and $\mathbf{B\tilde{x}} = \tilde{\mathbf{y}}$, then we can estimate

$$\|\tilde{\mathbf{x}} - \mathbf{x}\|_{\ell_2} \leq \|\mathbf{B}^{-1}\|_{\mathcal{L}(\ell_2)} \|\tilde{\mathbf{y}} - \mathbf{y}\|_{\ell_2}. \tag{7.2.23}$$

As a consequence, the ℓ_2 error for $\tilde{\mathbf{x}}_\varepsilon = \mathbf{SOLVE}[\mathbf{B}, \tilde{\mathbf{y}}, \varepsilon]$ is bounded by

$$\|\tilde{\mathbf{x}}_\varepsilon - \mathbf{x}\|_{\ell_2} \leq \varepsilon + \|\mathbf{B}^{-1}\|_{\mathcal{L}(\ell_2)} \|\tilde{\mathbf{y}} - \mathbf{y}\|_{\ell_2}. \tag{7.2.24}$$

Applying (7.2.24) to the approximate solution $\widetilde{\mathbf{D}_\alpha \mathbf{u}_1}$ of the first stage equation in the algorithm **ROW_INCREMENT**, we get the estimate

$$
\begin{aligned}
\|\widetilde{\mathbf{D}_\alpha \mathbf{u}_1} - \mathbf{D}_\alpha \mathbf{u}_1\|_{\ell_2} &\leq \varepsilon_1 + \|\mathbf{B}_\alpha^{-1}\| \|\tilde{\mathbf{r}}_{1,h} - \mathbf{r}_{1,h}\|_{\ell_2} \\
&\leq \varepsilon_1 + \|\mathbf{B}_\alpha^{-1}\| \big(\varepsilon_{1,1} + \varepsilon_{1,2} + T|\gamma_1|\varepsilon_{1,4}\big) \\
&= \theta \eta_1.
\end{aligned}
$$

Now assume that we are in the i–th for loop of the algorithm, $i \geq 2$, and the previous stage equations have been solved with accuracies $\theta \eta_j$, $j < i$. Then we can estimate in a completely analogous way

$$
\begin{aligned}
\|\tilde{\mathbf{r}}_{i,h} - \mathbf{r}_{i,h}\|_{\ell_2} &\leq \big\| \mathbf{D}_\alpha^{-1} \mathbf{D}_0 (\mathbf{B}_0(\mathbf{D}_0 \mathbf{u}^{(n)} + \mathbf{D}_0 \mathbf{D}_\alpha^{-1} \sum_{j=1}^{i-1} a_{i,j} \mathbf{D}_\alpha \mathbf{u}_j) - \mathbf{s}_{i,1}) \big\|_{\ell_2} \\
&\quad + \big\| \mathbf{D}_\alpha^{-1} \mathbf{D}_0 (\mathbf{D}^{-1} \langle f(t_n + \alpha_i h), \Psi \rangle^\top - \mathbf{s}_{i,2}) \big\|_{\ell_2} \\
&\quad + \big\| \mathbf{D}_\alpha^{-1} (\mathbf{G} \mathbf{D}_\alpha^{-1} \sum_{j=1}^{i-1} \tfrac{c_{i,j}}{h} \mathbf{D}_\alpha \mathbf{u}_j - \mathbf{s}_{i,3}) \big\|_{\ell_2} \\
&\quad + h|\gamma_i| \big\| \mathbf{D}_\alpha^{-1} \mathbf{D}_0 (\mathbf{D}_0^{-1} \langle f'(t_n), \Psi \rangle^\top - \mathbf{s}_{i,4}) \big\|_{\ell_2} \\
&\leq \varepsilon_{i,1} + \varepsilon_{i,2} + \varepsilon_{i,3} + T|\gamma_i|\varepsilon_{i,4} \\
&\quad + \big\| \mathbf{D}_\alpha^{-1} \mathbf{D}_0 \mathbf{B}_0 \mathbf{D}_0 \mathbf{D}_\alpha^{-1} \sum_{j=1}^{i-1} a_{i,j} (\widetilde{\mathbf{D}_\alpha \mathbf{u}_j} - \mathbf{D}_\alpha \mathbf{u}_j) \big\|_{\ell_2} \\
&\quad + \big\| \mathbf{D}_\alpha^{-1} \mathbf{G} \mathbf{D}_\alpha^{-1} \sum_{j=1}^{i-1} \tfrac{c_{i,j}}{h} (\widetilde{\mathbf{D}_\alpha \mathbf{u}_j} - \mathbf{D}_\alpha \mathbf{u}_j) \big\|_{\ell_2} \\
&\leq \varepsilon_{i,1} + \varepsilon_{i,2} + \varepsilon_{i,3} + T|\gamma_i|\varepsilon_{i,4} + \theta \sum_{j=1}^{i-1} (\|\mathbf{B}_0\| |a_{i,j}| + \|\mathbf{G}\| \tfrac{|c_{i,j}|}{h}) \eta_j,
\end{aligned}
$$

so that

$$
\|\widetilde{\mathbf{D}_\alpha \mathbf{u}_i} - \mathbf{D}_\alpha \mathbf{u}_i\|_{\ell_2} \leq \varepsilon_i + \|\mathbf{B}_\alpha^{-1}\| \|\tilde{\mathbf{r}}_{i,h} - \mathbf{r}_{i,h}\|_{\ell_2} \leq \theta \eta_i.
$$

For the stages $i \geq 2$, the tolerances ε_i and $\varepsilon_{i,\nu}$ have been chosen in such a way that $\eta_i \leq C \sum_{j=1}^{i-1} \eta_j$. By induction, one can hence prove the estimate $\eta_i \leq C(1+C)^{i-2}\eta_1$ for $i \geq 2$. Inserting the accuracies of the approximate stage solutions $\widetilde{\mathbf{D}_\alpha \mathbf{u}_i}$ into the final approximate increment shows that

$$
\begin{aligned}
\|\mathbf{D}_0 \check{\mathbf{u}}^{(n+1)} - \mathbf{D}_0 \mathbf{u}^{(n+1)}\|_{\ell_2} &\leq \big\| \mathbf{D}_0 \mathbf{D}_\alpha^{-1} \sum_{i=1}^{S} m_i (\widetilde{\mathbf{D}_\alpha \mathbf{u}_i} - \mathbf{D}_\alpha \mathbf{u}_i) \big\|_{\ell_2} \\
&\leq \sum_{i=1}^{S} |m_i| \|\widetilde{\mathbf{D}_\alpha \mathbf{u}_i} - \mathbf{D}_\alpha \mathbf{u}_i)\|_{\ell_2} \\
&\leq \theta \sum_{i=1}^{S} |m_i| \eta_i \\
&\leq \theta \big(|m_1| + C \sum_{i=2}^{S} |m_i| \sum_{j=1}^{i-1} (1+C)^{i-1}\big) \eta_1 \\
&= \theta \varepsilon.
\end{aligned}
$$

Since we have $\|\mathbf{D}_0 \check{\mathbf{u}}^{(n+1)} - \mathbf{D}_0 \check{\mathbf{u}}^{(n+1)}\|_{\ell_2} \leq (1 - \theta)\varepsilon$ by the choice of parameters and by the properties of the **COARSE** routine, the convergence claim (7.2.19) follows.

For the complexity estimate (7.2.20), observe first that the various tolerances are chosen in such a way that $\varepsilon_i, \varepsilon_{i,\nu} \gtrsim \varepsilon$. Hence, in order to prove (7.2.20), it suffices to control the ℓ_τ^w seminorms of the corresponding vector input parameters in the calls to the subroutines **APPLY**, **RHSF**, **RHSFP** and **SOLVE**. In fact, for the first stage equation, the compressibility of $\mathbf{B}_0$ implies that

$$
\# \operatorname{supp} \mathbf{s}_{1,1} \lesssim |\mathbf{B}_0 \mathbf{D}_0 \mathbf{u}^{(n)}|_{\ell_\tau^w}^{1/s} \varepsilon_{1,1}^{-1/s} \lesssim |\mathbf{B}_0 \mathbf{D}_0 \mathbf{u}^{(n)}|_{\ell_\tau^w}^{1/s} \varepsilon^{-1/s} \leq \tilde{C}^{1/s} \varepsilon^{-1/s},
$$

which by Lemma 4.8 yields $|\mathbf{s}_{1,1}|_{\ell^w_\tau} \lesssim |\mathbf{B}_0\mathbf{D}_0\mathbf{u}^{(n)}|_{\ell^w_\tau} \lesssim |\mathbf{D}_0\mathbf{u}^{(n)}|_{\ell^w_\tau} \leq \tilde{C}$. Using the properties of **APPLY**, we infer that the number of arithmetic operations to compute $\mathbf{s}_{1,1}$ is bounded by a multiple of $\# \operatorname{supp} \mathbf{s}_{1,1} \lesssim \tilde{C}^{1/s}\varepsilon^{-1/s}$. By the assumptions on the routine **RHSF**, it is

$$\# \operatorname{supp} \mathbf{s}_{1,2} \lesssim |\mathbf{D}_0^{-1}\langle f(t_n + \alpha_1 h), \Psi\rangle^\top|_{\ell^w_\tau}^{1/s}\varepsilon_{1,2}^{-1/s} \leq \tilde{C}^{1/s}\varepsilon^{-1/s},$$

so that again Lemma 4.8 implies $|\mathbf{s}_{1,2}|_{\ell^w_\tau} \lesssim \tilde{C}$ and the number of arithmetic operations to compute $\mathbf{s}_{1,2}$ is at most a multiple of $\# \operatorname{supp} \mathbf{s}_{1,2} \lesssim \tilde{C}^{1/s}\varepsilon^{-1/s}$. An analogous argument for $\mathbf{s}_{1,4}$ yields that finally

$$\# \operatorname{supp} \tilde{\mathbf{r}}_{1,h} \leq \# \operatorname{supp} \mathbf{s}_{1,1} + \# \operatorname{supp} \mathbf{s}_{1,2} + \# \operatorname{supp} \mathbf{s}_{1,4} \lesssim \tilde{C}^{1/s}\varepsilon^{-1/s},$$

so that $|\tilde{\mathbf{r}}_{1,h}|_{\ell^w_\tau} \lesssim \tilde{C}$ by Lemma 4.8, and $\tilde{\mathbf{r}}_{1,h}$ is computable with at most a multiple of $\tilde{C}^{1/s}\varepsilon^{-1/s}$ arithmetic operations. As an output of **SOLVE**, the first approximate stage solution $\widetilde{\mathbf{D}_\alpha\mathbf{u}_1}$ fulfills the estimate

$$\# \operatorname{supp} \widetilde{\mathbf{D}_\alpha\mathbf{u}_1} \lesssim |\mathbf{B}_\alpha^{-1}\tilde{\mathbf{r}}_{1,h}|_{\ell^w_\tau}^{1/s}\varepsilon_1^{-1/s} \lesssim \tilde{C}^{1/s}\varepsilon^{-1/s},$$

and hence $|\widetilde{\mathbf{D}_\alpha\mathbf{u}_1}|_{\ell^w_\tau} \lesssim \tilde{C}$ by Lemma 4.8, the number of arithmetic operations to compute $\widetilde{\mathbf{D}_\alpha\mathbf{u}_1}$ being bounded by a constant times $\tilde{C}^{1/s}\varepsilon^{-1/s}$.

Now assume that we are in the i–th for loop, $i \geq 2$, and the previous stages have been computed with $\# \operatorname{supp} \widetilde{\mathbf{D}_\alpha\mathbf{u}_j} \lesssim \tilde{C}^{1/s}\varepsilon^{-1/s}$ and $|\widetilde{\mathbf{D}_\alpha\mathbf{u}_j}|_{\ell^w_\tau} \lesssim \tilde{C}$. Then we can infer that also

$$|\mathbf{D}_0\mathbf{u}^{(n)} + \mathbf{D}_0\mathbf{D}_\alpha^{-1}\sum_{j=1}^{i-1} a_{i,j}\widetilde{\mathbf{D}_\alpha\mathbf{u}_j}|_{\ell^w_\tau} \lesssim \tilde{C},$$

and $\mathbf{D}_0\mathbf{u}^{(n)} + \mathbf{D}_0\mathbf{D}_\alpha^{-1}\sum_{j=1}^{i-1} a_{i,j}\widetilde{\mathbf{D}_\alpha\mathbf{u}_j}$ can be computed with at most a constant times $\tilde{C}^{1/s}\varepsilon^{-1/s}$ operations. By the properties of **APPLY** and the compressibility of $\mathbf{B}_0$, we know that

$$\# \operatorname{supp} \mathbf{s}_{i,1} \lesssim \left|\mathbf{B}_0(\mathbf{D}_0\mathbf{u}^{(n)} + \mathbf{D}_0\mathbf{D}_\alpha^{-1}\sum_{j=1}^{i-1} a_{i,j}\widetilde{\mathbf{D}_\alpha\mathbf{u}_j})\right|_{\ell^w_\tau}^{1/s}\varepsilon_{i,1}^{-1/s} \lesssim \tilde{C}^{1/s}\varepsilon^{-1/s},$$

so that Lemma 4.8 yields $|\mathbf{s}_{i,1}|_{\ell^w_\tau} \lesssim \tilde{C}$. Moreover, the number of arithmetic operations to compute $\mathbf{s}_{i,1}$ is bounded by a multiple of $\# \operatorname{supp} \mathbf{s}_{i,1} \lesssim \tilde{C}^{1/s}\varepsilon^{-1/s}$. An analogous argument for the Gramian $\mathbf{G}$ shows that

$$\# \operatorname{supp} \mathbf{s}_{i,3} \lesssim |\mathbf{G}\mathbf{D}_\alpha^{-1}\sum_{j=1}^{i-1} \frac{c_{i,j}}{h}\widetilde{\mathbf{D}_\alpha\mathbf{u}_j}|_{\ell^w_\tau}^{1/s}\varepsilon_{i,3}^{-1/s} \lesssim \tilde{C}^{1/s}\varepsilon^{-1/s},$$

with $|\mathbf{s}_{i,3}|_{\ell^w_\tau} \lesssim \tilde{C}$ and the number of arithmetic operations to compute $\mathbf{s}_{i,3}$ being bounded by a constant times $\# \operatorname{supp} \mathbf{s}_{i,3} \lesssim \tilde{C}^{1/s}\varepsilon^{-1/s}$. Using again the assumptions on **RHSF** and on **RHSFP** like in the first stage equation, we end up with an approximate i–th right–hand side $\tilde{\mathbf{r}}_{i,h}$ with $|\tilde{\mathbf{r}}_{i,h}|_{\ell^w_\tau} \lesssim \tilde{C}$. By the properties of **SOLVE**, we can deduce that

$$\# \operatorname{supp} \widetilde{\mathbf{D}_\alpha\mathbf{u}_i} \lesssim |\mathbf{B}_\alpha^{-1}\tilde{\mathbf{r}}_{i,h}|_{\ell^w_\tau}^{1/s}\varepsilon_i^{-1/s} \lesssim \tilde{C}^{1/s}\varepsilon^{-1/s},$$

and hence $|\widetilde{\mathbf{D}_\alpha\mathbf{u}_i}|_{\ell^w_\tau} \lesssim \tilde{C}$ by Lemma 4.8. The number of arithmetic operations to compute $\widetilde{\mathbf{D}_\alpha\mathbf{u}_i}$ is bounded by at most a multiple of $\tilde{C}^{1/s}\varepsilon^{-1/s}$.

The final claim (7.2.20) follows by the properties of the **COARSE** routine. Namely, knowing that $\|\mathbf{D}_0 \check{\mathbf{u}}^{(n+1)} - \mathbf{D}_0 \mathbf{u}^{(n+1)}\|_{\ell_2} \leq \theta\varepsilon$ from the convergence proof, we can invoke (4.4.3) to infer that

$$\# \operatorname{supp} \mathbf{D}_0 \tilde{\mathbf{u}}^{(n+1)} \lesssim |\mathbf{D}_0 \mathbf{u}^{(n+1)}|_{\ell_\tau^w}^{1/s} \varepsilon^{-1/s}.$$

By the preceding arguments, the computation of $\mathbf{D}_0 \tilde{\mathbf{u}}^{(n+1)}$ involves only a constant times $\tilde{C}^{1/s} \varepsilon^{-1/s}$ arithmetic operations. The proof is complete. $\qquad\square$

Remark 7.11. *In practice, of course, one will not implement Algorithm 7.8 exactly as it is printed here. This is mainly due to the fact that the worst case estimates done in the convergence proof are very pessimistic, which may result in a bad quantitative response of the algorithm. Instead, in the numerical experiments of Chapter 8, we set most of the constants in the algorithm to reasonable values and rely on the fact that* **ROW_INCREMENT** *will still yield approximations with* $\|\mathbf{D}_0 \tilde{\mathbf{u}}^{(n+1)} - \mathbf{D}_0 \mathbf{u}^{(n+1)}\|_{\ell_2} \leq C'\varepsilon$, *with a moderate constant C'. This approach is common in adaptive wavelet methods, see also* [7, 52, 103].

Chapter 8

Numerical Experiments

This chapter is devoted to the numerical validation of the convergence and complexity results for the adaptive wavelet schemes introduced in Chapter 7. In particular, we shall study the temporal and spatial convergence for several one– and two–dimensional examples. We will also address auxiliary algorithms that are used in the course of the experiments.

8.1 Design of the Experiments

In the experiments, we shall study homogeneous and inhomogeneous parabolic problems of the form (0.0.20) and (0.0.21), respectively. The setting will be restricted to that of the heat equation, i.e., $A = \Delta_\Omega^D$ is the Dirichlet Laplacian operator over the domain Ω. We consider test cases on the time interval $[0, 1]$ where the initial value and the driving term f are either smooth or non–smooth. Here spatial smoothness is measured in the scale of spaces $D(A^k)$. The test examples are consecutively numbered, and a rough overview of them is given in Table 8.1. More details are addressed in the following subsections.

d	Ω	u_0	f	nr.
1	$(0, 1)$	smooth	0	1
		non–smooth		2
		0	temp. and spatially smooth, tensor prod.	3
			temp. smooth, spatially non–smooth	4
			temp. and spatially non–smooth	5
		smooth	temp. and spatially smooth, no tensor prod.	6
2	L–domain	0	temp. and spatially smooth	7
			temp. and spatially non–smooth	8

Table 8.1: Survey of the parabolic test examples.

8.1.1   1D Examples

Homogeneous Problems

For a homogeneous parabolic problem (0.0.20), we know from the previous discussion that the exact solution $u(t)$ is given by the semigroup action $u(t) = e^{tA}u_0$ on the initial value u_0. Hence, for $\Omega = (0, 1)$, a reference solution $u(t)$ at time $t > 0$ can be computed via the series expansion (6.3.4). We shall essentially use the same test examples as in [130]. More precisely, we will consider both a smooth and a nonsmooth initial value u_0 with $\|u_0\|_{L_2(0,1)} = 1$.

In the test example 1, we choose the initial value

$$u_0(x) = \sqrt{\frac{2}{5}}\big(\sin(\pi x) + 2\sin(2\pi x)\big), \quad x \in (0, 1), \tag{8.1.1}$$

so that u_0 and hence $u(t)$ are non–symmetric as functions in the spatial variable. We have $u_0 \in D(A^k)$ for any power $k \in \mathbb{N}$ since u_0 is a finite linear combination of eigenfunctions of the Dirichlet Laplacian $A = \frac{d^2}{dx^2}$. Consequently, the solution u has an arbitrarily large temporal smoothness, both as a mapping into $L_2(\Omega)$ and into $D(A)$. Moreover, in the sine series expansion of u_0 and in (6.3.4), all but two entries are zero, so that the reference solution $u(t)$ is a finite sine sum and can hence be evaluated up to machine precision in a stable way by the Goertzel–Reinsch algorithm, see [68] for details. A plot of u_0 and of the corresponding solution u is given in Figure 8.1.

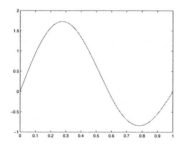 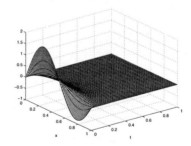

Figure 8.1: Test example 1: Initial value u_0 (*left*) and the corresponding reference solution u (*right*).

In the test example 2, we would like to choose the non–smooth initial value

$$u_0(x) = \sqrt{2}\chi_{[\frac{1}{4}, \frac{3}{4}]}(x), \quad x \in (0, 1). \tag{8.1.2}$$

Here it is only $u_0 \in H^{1/2-\epsilon}(0, 1)$ for all $\epsilon > 0$, so that $u_0 \notin D(A)$. However, u_0 is contained in the domain of certain fractional powers of A. Since A is self–adjoint and $D(A) = H^2(\Omega) \cap H_0^1(\Omega)$, we can infer from (6.1.23) that

$$D(A^s) = [L_2(\Omega), H^2(\Omega) \cap H_0^1(\Omega)]_{s,2}, \quad s \in (0, 1). \tag{8.1.3}$$

This scale of spaces has been studied, e.g., in [80]. For $s < \frac{1}{4}$, it was proved in loc. cit. that $D(A^s) = H^{2s}(\Omega)$, so that $u_0 \in D(A^{1/4-\epsilon})$ for all $\epsilon > 0$. The latter fact can also be derived by analytically computing the expansion coefficients with respect to the orthonormal eigenfunctions $\{v_k\}_{k\in\mathbb{N}}$ from (6.3.1)

$$\langle u_0, v_k \rangle_{L_2(0,1)} = \frac{4}{k\pi} \sin(k\pi/2) \sin(k\pi/4), \quad k \in \mathbb{N}, \tag{8.1.4}$$

and by using the *Picard criterion*

$$D(A^s) = \left\{ f \in L_2(0,1) : \sum_{k\in\mathbb{N}} \lambda_k^{2s} |\langle f, v_k \rangle_{L_2(0,1)}|^2 < \infty \right\}, \quad s > 0. \tag{8.1.5}$$

Since the sine series expansion (6.3.4) of u_0 and of the corresponding solution u involves infinitely many nontrivial terms, we cannot evaluate the reference solution at time t with machine precision. Instead, we truncate the sine series expansion of u_0 at a fixed maximal index K and compute an approximate initial value

$$\tilde{u}_0 = \tilde{u}_0(\cdot; K) = \sum_{0 \le k \le K} \langle u_0, v_k \rangle v_k.$$

From (8.1.4), we infer that $\|\tilde{u}_0 - u_0\|_{L_2(0,1)} = \mathcal{O}(K^{-1/2})$ as K tends to infinity. In the numerical experiments, we choose the truncation parameter $K = 20000$, which amounts to $\|\tilde{u}_0 - u_0\|_{L_2(0,1)} \le 10^{-2}$. Temporal and spatial errors are computed against the solution u corresponding to the truncated initial value $\tilde{u}_0$ which shall hence be also denoted by u_0 in the sequel. A plot of u_0 and of the corresponding solution u can be seen in Figure 8.2. It becomes visible that the non–smooth initial value u_0 is smoothed rapidly by the semigroup at the beginning of the time interval. Since the initial value u_0 is non–smooth, the corresponding solution u has a limited temporal Hölder smoothness due to Theorem 6.8, which will be of importance for the numerical experiments.

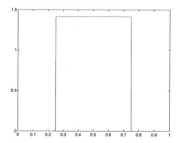 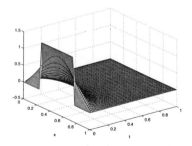

Figure 8.2: Test example 2: Initial value u_0 (*left*) and the corresponding reference solution u (*right*).

Inhomogeneous Problems

We consider inhomogeneous 1D problems of the form (0.0.21), i.e., we choose $u_0 = 0$ and several driving terms f of different temporal and spatial smoothness each. Consequently, the exact solution u is given as the convolution integral (6.2.1). For the special case that $f(s) \equiv f_0$ is temporally constant, u has the series expansion

$$u(t) = \int_0^t e^{(t-s)A} f(s)\, ds = \sum_{k \in \mathbb{N}} \langle f_0, v_k \rangle \int_0^t e^{-(t-s)k^2\pi^2} v_k\, ds = \sum_{k \in \mathbb{N}} \frac{1 - e^{-k^2\pi^2 t}}{k^2\pi^2} \langle f_0, v_k \rangle v_k.$$

(8.1.6)

The test example 3 shall be of this type, where we consider the driving term

$$f(t,x) = \pi^2 \sin(\pi x), \quad t \in [0,1], \quad x \in \Omega, \tag{8.1.7}$$

which is a multiple of the eigenfunction v_1 and corresponds to the exact solution

$$u(t,x) = (1 - e^{-\pi^2 t}) \sin(\pi x), \quad t \in [0,1], \quad x \in \Omega. \tag{8.1.8}$$

f is arbitrarily smooth both in time and in space. Note that part (iii) of Theorem 6.13 is not applicable since $f(0) \neq 0$. Nevertheless, u is also temporally and spatially smooth by definition. A plot of the driving term f and the corresponding solution u can be found in Figure 8.3.

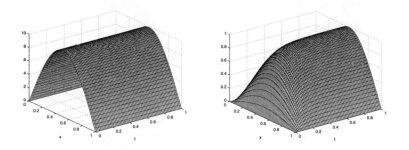

Figure 8.3: Test example 3: Driving term f (*left*) and the corresponding reference solution u (*right*).

In the test example 4, we choose the driving term

$$f(t,x) = 1, \quad t \geq 0, \quad x \in \Omega. \tag{8.1.9}$$

Although f is arbitrarily smooth in time, the function $f(t,\cdot) = f_0$ does not fulfill the spatial boundary conditions. More precisely, since f_0 is (piecewise) constant, we only have $f_0 \in D(A^{1/4-\epsilon})$ for all $\epsilon > 0$ similar to the initial value u_0 in test example 2. Consequently, the exact solution $u(t,\cdot)$ at t is only contained in $D(A^{5/4-\epsilon})$. By Theorem 6.12, we derive no more temporal regularity of u than $u \in C^\alpha([0,T]; L_2)$ for all $0 < \alpha < 1$. We can compute u by the series expansion 8.1.6, truncated at some maximal eigenvalue, see also Figure 8.4.

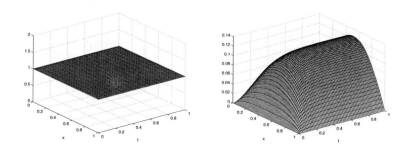

Figure 8.4: Test example 4: Driving term f (*left*) and the corresponding reference solution u (*right*).

In order to derive an example where f is non–smooth both in time and space, we consider driving terms of the form $f(t) = \chi_{[0,a)}(t)f_0$, so that it is only $f \in H^{1/2-\epsilon}(0, T; L_2(\Omega))$ for all $\epsilon > 0$. There, analogously to (8.1.6), the exact solution u can again be computed as a series

$$u(t) = \sum_{k \in \mathbb{N}} \frac{\min\{1, e^{-k^2\pi^2(t-a)}\} - e^{-k^2\pi^2 t}}{k^2\pi^2} \langle f_0, v_k \rangle v_k, \quad t \geq 0. \tag{8.1.10}$$

As test example 5, we choose a right–hand side of the aforementioned type

$$f(t, x) = \sqrt{2}\chi_{[0,\frac{1}{2})}(t)\chi_{[\frac{1}{4},\frac{3}{4})}(x), \quad t \in [0, 1], \quad x \in \Omega. \tag{8.1.11}$$

For fixed $t \in [0, 1]$, we have $f(t, \cdot) \in D(A^{1/4-\epsilon})$ for all $\epsilon > 0$, and $t \mapsto f(t, \cdot)$ has a discontinuity at $t = \frac{1}{2}$. A plot of f and the corresponding solution u can be found in Figure 8.5.

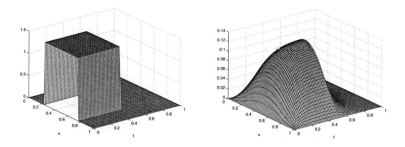

Figure 8.5: Test example 5: Driving term f (*left*) and the corresponding reference solution u (*right*).

Finally, as test example 6, we choose the exact solution u in such a way that f can *not* be written as a tensor product of two univariate functions. Precisely, we assume that u is a moving Gaussian

$$u(t, x) = e^{-300(x - 0.6 + 0.2t)^2}, \quad t \in [0, 1], \quad x \in \Omega, \tag{8.1.12}$$

so that $u(t, \cdot)$ fulfills the homogeneous boundary conditions at least numerically. The driving term is chosen as $f(t, x) = u_t(t, x) - u_{xx}(t, x)$. We note that for the given solution u, the driving term f has a nontrivial temporal derivative, in contrast to the other one–dimensional examples. A plot of f and the corresponding solution u is given in Figure 8.6.

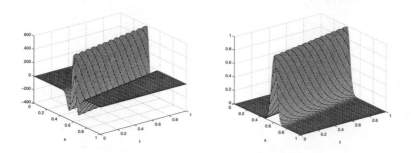

Figure 8.6: Test example 6: Driving term f (*left*) and the corresponding reference solution u (*right*).

8.1.2 2D Examples

Concerning the numerical examples on the L–shaped domain $\Omega = (-1, 1)^2 \setminus [0, 1)^2$, we shall only study inhomogeneous problems of the form (0.0.20), where $u_0 = 0$. This is mainly due to the fact that a complete eigensystem for the Dirichlet Laplacian is not available in this situation, making the computation of reference solutions for homogeneous parabolic problems difficult.

The test example 7 is chosen in analogy to example 3, where the driving term

$$f(t, x, y) = 2\pi^2 \sin(\pi x) \sin(\pi y), \quad t \in [0, 1], \quad (x, y)^\top \in \Omega \tag{8.1.13}$$

is temporally constant. f is spatially smooth with $f(t, \cdot, \cdot) \in D(A^k)$ for all $k \in \mathbb{N}$ since $f(t, \cdot, \cdot)$ is an eigenfunction of the Dirichlet Laplacian Δ_Ω^D. The exact solution u is the tensor product

$$u(t, x, y) = (1 - e^{-2\pi^2 t}) \sin(\pi x) \sin(\pi y), \quad t \in [0, 1], \quad (x, y)^\top \in \Omega \tag{8.1.14}$$

which is also temporally and spatially smooth. In Figure 8.7, we give a plot of u at the time $t = 1$.

Finally, test example 8 shall be designed in such a way that the reference solution u is neither temporally nor spatially smooth. We choose the function

$$u(t, x, y) = t^{3/4} r^{2/3} \sin\left(\frac{2\theta}{3\pi}\right)(1 - x^2)(1 - y^2), \quad t \in [0, 1], \quad (x, y)^\top \in \Omega, \tag{8.1.15}$$

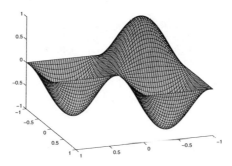

Figure 8.7: Test example 7: Reference solution $u(t, x, y)$ at $t = 1$

which has also been used in the experiments of [83]. Here $(r, \theta) = (\sqrt{x^2 + y^2}, \arctan \frac{y}{x})$ are the polar coordinates with respect to the reentrant corner at $(x, y)^{\top} = (0, 0)^{\top}$. Note that u has a temporal singularity at $t = 0$ and a spatial singularity at the origin. Precisely, $u(t, \cdot, \cdot)$ is contained only in $H^{5/3}(\Omega)$ due to the behavior at the reentrant corner. Contrary to that, we have $u(t, \cdot, \cdot) \in B_{\tau}^{2s+1}(L_{\tau}(\Omega))$ for *all* $s \geq 0$, $\tau = (s + \frac{1}{2})^{-1}$, since u is the pointwise product of the analytic function $(x, y)^{\top} \mapsto (1 - x^2)(1 - y^2)$ and a function of arbitrary high Besov regularity. Figure 8.8 shows a plot of u at the time $t = 1$. The right–hand side is chosen according

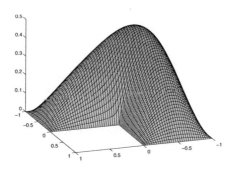

Figure 8.8: Test example 8: Reference solution $u(t, x, y)$ at $t = 1$

to $f(t, x, y) = u_t(t, x, y) - \Delta u(t, x, y)$ which has a nontrivial temporal derivative, similar to the driving term in example 6. Both functions f and f_t are unbounded for $t \to 0$ and they do no longer fulfill the spatial Dirichlet boundary conditions.

8.1.3   More Details on the Temporal Discretization

For the temporal discretization, we select several appropriate ROW–methods with different number of stages S and different orders of consistency p. The concrete choice of methods is guided by the suggestions from [108]. All of the schemes are both $A(\pi/2)$– and L–stable, and they are listed in Table 8.2.

name	reference	S	p	further properties
ROS2	[16]	2	2	
ROWDA3	[127]	3	3	index 1 DAEs
ROS3Pw	[3]	3	3	index 1, PDEs, W–method
RODASP	[132]	6	4	index 1 DAEs, PDEs, stiffly accurate

Table 8.2: Survey of the ROW methods used in the numerical experiments.

The ROS2 time integrator from [16] is one of the two possible second–order ROW–schemes with two stages that are also L–stable. Note that in [103], the alternative variant was used as a second order L–stable benchmark solver. However, the ROS2 solver may have better stability properties since its stability function $R(z)$ is positive for all $z \leq 0$, which is not the case for the alternative second order scheme. We point at the discussion in Section 3.2 of [16] here.

As a principal third–order scheme, we choose the three–stage solver ROWDA3 from [127]. Although it has not been specifically designed for applications in partial differential equations, the scheme generally performs well in the case of linear parabolic equations with homogeneous Dirichlet boundary conditions due to Theorem 7.4. In test example 6 we shall see that steep gradients of the exact solution may lead to inferior convergence behaviour of ROWDA3 for small stepsizes. Hence, for comparison, we add results for the more sophisticated third–order integrator ROS3Pw from [3] in this case. The latter scheme fulfills additional stability properties that make it suitable also for the case of W–methods where J and g are only approximations of A and $f'(t_n)$, respectively. Apart from the schemes ROWDA3 and ROS3Pw, we have tested further third–order methods, e.g., ROS3P [109] or ROSI2PW [3]. However, for the class of linear parabolic problems we are interested in, the corresponding numerical results do not differ significantly from those obtained with the chosen. This is essentially due to the fact that the order reduction for ROW methods applied to discretizations of linear homogeneous Dirichlet problems is not as severe as it would be the case for inhomogeneous or Neumann problems. When it comes to applications of ROW–methods to nonlinear parabolic problems, specially tailored schemes like ROS3P would be preferable.

The fourth–order integrator RODASP developed in [132] complements the choice of Rosenbrock methods. The scheme is specifically designed for the application to PDEs, i.e, it fulfills the additional algebraic order conditions from Theorem 7.4. Since the RODASP scheme is quite expensive due to the high number of stages, we have used it only for the one–dimensional examples.

8.2 Temporal Convergence for Constant Stepsizes

In a first step, we verify the temporal convergence results stated in Subsection 7.1.2 for the one–dimensional problems 1 to 6 and the special case of a wavelet discretization in space. In a finite element context, computations of this type have been done in [107]. We redo the same steps in order to verify the integrity of the implementation.

For the validation of the asserted temporal convergence orders in one space dimension, we apply the mentioned ROW methods with a constant temporal stepsize h to the test examples 1 to 6 on the time interval $[0, T]$. Due to the parabolic smoothing, the exact solution of a homogeneous problem decays exponentially fast to zero. Therefore, we set the value of T in the numerical experiments to $T = 0.1$ for the problems 1 and 2, and to $T = 1$ in the inhomogeneous problems 3 to 6. In order to suppress the additional spatial discretization error to an insignificant level, we use a maximal refinement level $j_{\max} = 15$ and quadratic spline wavelet bases $(m = \tilde{m} = 3)$ on the interval from [126]. Since the exact solutions u are known in each testcase, we can explicitly compute the L_2 errors

$$e_h := \max_{k=0,\dots,T/h} \|u(kh) - u^{(k)}\|_{L_2(\Omega)} \tag{8.2.1}$$

for a constant temporal stepsize h by applying a suitable quadrature rule. Under the assumption that the discretization error behaves like $e_h = ch^p$, the values

$$p_{\mathrm{num}} := \log_2 \frac{e_h}{e_{h/2}} \tag{8.2.2}$$

may serve as numerical estimators of the temporal convergence order p.

8.2.1 Homogeneous Problems

	ROS2		ROWDA3		RODASP	
h	e_h	p_{num}	e_h	p_{num}	e_h	p_{num}
2^{-2}	$8.65e-2$		$5.57e-3$		$2.83e-4$	
2^{-3}	$3.94e-2$	1.13	$8.07e-4$	2.79	$1.69e-5$	4.06
2^{-4}	$1.50e-2$	1.39	$1.12e-4$	2.85	$1.04e-6$	4.02
2^{-5}	$4.92e-3$	1.61	$1.49e-5$	2.91	$6.48e-8$	4.01
2^{-6}	$1.44e-3$	1.77	$1.93e-6$	2.95	$4.06e-9$	4.00
2^{-7}	$3.92e-4$	1.88	$2.46e-7$	2.97	$2.72e-10$	3.90
2^{-8}	$1.03e-4$	1.93	$3.10e-8$	2.99		
2^{-9}	$2.63e-5$	1.97	$3.89e-9$	2.99		
2^{-10}	$6.64e-6$	1.98	$4.87e-10$	3.00		

Table 8.3: Test example 1: L_2 convergence for constant stepsizes

In Table 8.3, we report the results for the test example 1. The predicted convergence rates for homogeneous problems with a smooth initial value u_0 can be observed for all of the three time integrators considered. Of course the order estimator (8.2.2)

fails for very small values of h since then, the spatial truncation error dominates the temporal error and the assumption $e_h = ch^p$ is no longer valid. For this reason, we omit some of the results for RODASP.

In test example 2, we expect a limitation of the numerical convergence rates due to the non–smooth initial value u_0. As stated above, the exact solution u is contained only in $C^{0.25-\epsilon}([0,T]; L_2)$. Table 8.4 reveals that all time integrators yield approximations with $e_h \approx h^{0.25}$ only which matches the regularity of u. For very small stepsizes $h \leq 2^{-9}$, the spatial discretization error spoils the observable convergence orders for all time integrators, since the chosen maximal discretization level $j_{\max} = 15$ is not sufficient to approximate the non–smooth functions u_0 and $u(t, \cdot)$ for $t \approx 0$ with an L_2 error smaller than 10^{-2}.

h	ROS2 e_h	p_{num}	ROWDA3 e_h	p_{num}	RODASP e_h	p_{num}
2^{-2}	$5.62e-2$		$3.26e-2$		$3.55e-2$	
2^{-3}	$4.84e-2$	0.22	$2.75e-2$	0.24	$2.98e-2$	0.25
2^{-4}	$4.03e-2$	0.26	$2.31e-2$	0.25	$2.50e-2$	0.25
2^{-5}	$3.39e-2$	0.25	$1.94e-2$	0.25	$2.09e-2$	0.26
2^{-6}	$2.85e-2$	0.25	$1.62e-2$	0.26	$1.71e-2$	0.29
2^{-7}	$2.39e-2$	0.25	$1.34e-2$	0.28	$1.31e-2$	0.38
2^{-8}	$2.00e-2$	0.26	$1.05e-2$	0.35	$8.26e-3$	0.67
2^{-9}	$1.65e-2$	0.28	$7.06e-3$	0.57	$2.81e-3$	1.55
2^{-10}	$1.28e-2$	0.36	$2.84e-3$	1.31	$2.69e-3$	0.07

Table 8.4: Test example 2: L_2 convergence for constant stepsizes

8.2.2　Inhomogeneous Problems

For the inhomogeneous test problem 3, having a smooth right–hand side, the theoretical considerations from Subsection 8.1.1 predict the classical convergence orders 2, 3 and 4 for all of the considered time integrators. This can be validated numerically, see Table 8.5.

h	ROS2 e_h	p_{num}	ROWDA3 e_h	p_{num}	RODASP e_h	p_{num}
2^{-2}	$1.21e-1$		$3.53e-2$		$6.41e-3$	
2^{-3}	$8.57e-2$	0.50	$8.00e-3$	2.14	$5.50e-4$	3.54
2^{-4}	$4.07e-2$	1.07	$1.16e-3$	2.78	$3.22e-5$	4.09
2^{-5}	$1.64e-2$	1.31	$1.68e-4$	2.79	$2.01e-6$	4.00
2^{-6}	$5.64e-3$	1.54	$2.26e-5$	2.89	$1.25e-7$	4.01
2^{-7}	$1.71e-3$	1.72	$2.95e-6$	2.94	$7.82e-9$	4.00
2^{-8}	$4.74e-4$	1.84	$3.77e-7$	2.97	$5.03e-10$	3.96
2^{-9}	$1.25e-4$	1.92	$4.77e-8$	2.98		
2^{-10}	$3.22e-5$	1.96	$6.00e-9$	2.99		

Table 8.5: Test example 3: L_2 convergence for constant stepsizes

In the test examples 4 and 5, the driving term f and the corresponding solution u are at least spatially non–smooth. In both cases, f is only mapping into $D(A^{1/4-\epsilon})$, so that the solution $u(t, \cdot)$ at t is only contained in $D(A^{5/4-\epsilon})$. By Theorem 6.12, we have at least that $u \in C^\alpha([0, T]; L_2)$ for all $0 < \alpha < 1$, so that a convergence rate greater or equal to 1 can be expected in both examples. The results for example 4 are given in Table 8.6. Obviously, all of the considered schemes exhibit the L_2 convergence rate $e_h \approx h^{1.25}$, in accordance with the theory.

	ROS2		ROWDA3		RODASP	
h	e_h	p_{num}	e_h	p_{num}	e_h	p_{num}
2^{-2}	$1.56e{-}02$		$4.56e{-}03$		$1.01e{-}03$	
2^{-3}	$1.11e{-}02$	0.50	$1.12e{-}03$	2.03	$5.16e{-}04$	0.97
2^{-4}	$5.25e{-}03$	1.07	$4.35e{-}04$	1.36	$2.58e{-}04$	1.00
2^{-5}	$2.12e{-}03$	1.31	$2.27e{-}04$	0.94	$1.02e{-}04$	1.33
2^{-6}	$7.28e{-}04$	1.54	$9.24e{-}05$	1.30	$4.34e{-}05$	1.24
2^{-7}	$2.45e{-}04$	1.57	$3.89e{-}05$	1.25	$1.82e{-}05$	1.25
2^{-8}	$1.08e{-}04$	1.19	$1.63e{-}05$	1.25	$7.66e{-}06$	1.25
2^{-9}	$4.49e{-}05$	1.26	$6.87e{-}06$	1.25	$3.22e{-}06$	1.25
2^{-10}	$1.89e{-}05$	1.25	$2.89e{-}06$	1.25	$1.35e{-}06$	1.25

Table 8.6: Test example 4: L_2 convergence for constant stepsizes

	ROS2		ROWDA3		RODASP	
h	e_h	p_{num}	e_h	p_{num}	e_h	p_{num}
2^{-2}	$2.69e{-}02$		$4.56e{-}03$		$2.59e{-}02$	
2^{-3}	$2.01e{-}02$	0.42	$1.12e{-}03$	2.03	$1.46e{-}02$	0.82
2^{-4}	$1.44e{-}02$	0.48	$4.35e{-}04$	1.36	$7.85e{-}03$	0.90
2^{-5}	$9.49e{-}03$	0.60	$2.27e{-}04$	0.94	$4.09e{-}03$	0.94
2^{-6}	$5.69e{-}03$	0.74	$9.24e{-}05$	1.30	$2.10e{-}03$	0.96
2^{-7}	$3.18e{-}03$	0.84	$3.89e{-}05$	1.25	$1.07e{-}03$	0.97
2^{-8}	$1.70e{-}03$	0.90	$1.63e{-}05$	1.25	$5.42e{-}04$	0.98
2^{-9}	$8.89e{-}04$	0.94	$6.88e{-}06$	1.25	$2.73e{-}04$	0.99
2^{-10}	$4.57e{-}04$	0.96	$2.91e{-}06$	1.24	$1.37e{-}04$	0.99

Table 8.7: Test example 5: L_2 convergence for constant stepsizes

The results for example 5, as given in Table 8.7, are comparable to those for example 4. Here the driving term f has a discontinuity at $t = \frac{1}{2}$, so that u has a limited temporal Hölder smoothness. We observe the minimally expected convergence rates of $p_{num} = 1$ for the schemes ROS2 and RODASP, whereas the third–order integrator ROWDA3 exhibits a slightly better performance $p_{num} = 1.25$. Add to the better convergence rate, the attained minimal error for ROWDA3 is much smaller when compared to the schemes ROS2 and even RODASP. This effect can be explained as follows. For a given time step t_n with stepsize h, a Rosenbrock scheme evaluates the right–hand side at the points $t_n + \alpha_i h$. In both integrators ROS2 and RODASP, there is at least one coefficient $\alpha_i = 1$. More precisely, we have $(\alpha_i)_{i=1}^2 = (0, 1)$ for ROS2 and $(\alpha_i)_{i=1}^6 = (0, 0.75, 0.21, 0.63, 1, 1)$ for RODASP. So for the particular choice of equidistant dyadic stepsizes, both schemes will evaluate f at the discontinuity $t = \frac{1}{2}$ not only when $t_n = \frac{1}{2}$ but also in the case $t_n + h = \frac{1}{2}$. The latter situation is more severe since for small stepsizes, we end up with a steep numerical slope of f despite the fact that $f' = 0$. This effect can not happen for ROWDA3, where we have $(\alpha_i)_{i=1}^3 = (0, 0.7, 0.7)$. Here a point evaluation of f at $t = \frac{1}{2}$ only occurs at the beginning of a time step, which does not have such dramatical consequences.

In the test example 6, both the driving term f and the exact solution u are temporally and spatially smooth, so that we should observe the classical integer convergence orders in the numerical experiments. As seen in Table 8.8, the second order scheme ROS2 indeed converges with a rate $p_{\text{num}} \geq 2$, as expected. For the higher order schemes, however, the numerical rates slightly deviate from the theoretical ones. The ROWDA3 integrator shows the rate 3 for moderate stepsizes h only, but for small stepsizes the rate decreases to values close to 2. Analogously, the classical fourth order of RODASP is reduced to $p_{\text{num}} > 3$. We suspect that the slower convergence in both cases stems from the fact that the temporal derivatives of f are highly oscillatory in space so that the approximate evaluation of the coefficient array $\langle f'(t_n), \Psi \rangle^\top$ leads to additional spatial errors per increment step that are no longer negligible. Hence in test example 6, ROW–methods behave like W–methods where additional order conditions have to be fulfilled to sustain the classical convergence orders. As a complement, we thus add computational results for the scheme ROS3Pw in Table 8.8. Here the classical order $p = 3$ can indeed be observed in practice. We have also made test computations with other nonseparable driving terms f the temporal derivatives of which were less oscillating. In these cases the scheme ROWDA3 also shows a third–order convergence.

	ROS2		ROWDA3		ROS3Pw		RODASP	
h	e_h	p_{num}	e_h	p_{num}	e_h	p_{num}	e_h	p_{num}
2^{-2}	$1.10e{-}01$		$3.01e{-}02$		$4.29e{-}02$		$1.61e{-}05$	
2^{-3}	$2.96e{-}02$	1.90	$3.99e{-}03$	2.91	$3.91e{-}03$	3.46	$1.60e{-}06$	3.33
2^{-4}	$7.43e{-}03$	1.99	$5.14e{-}04$	2.96	$5.07e{-}04$	2.95	$2.06e{-}07$	2.96
2^{-5}	$1.81e{-}03$	2.03	$8.49e{-}05$	2.60	$6.43e{-}05$	2.98	$2.92e{-}08$	2.82
2^{-6}	$4.29e{-}04$	2.08	$2.35e{-}05$	1.85	$8.13e{-}06$	2.98	$3.99e{-}09$	2.87
2^{-7}	$9.66e{-}05$	2.15	$7.36e{-}06$	1.68	$1.03e{-}06$	2.98	$4.86e{-}10$	3.04
2^{-8}	$2.02e{-}05$	2.26	$2.04e{-}06$	1.85	$1.29e{-}07$	2.99	$5.04e{-}11$	3.27
2^{-9}	$4.20e{-}06$	2.26	$4.84e{-}07$	2.08	$1.58e{-}08$	3.03	$4.93e{-}12$	3.35
2^{-10}	$1.34e{-}06$	1.65	$1.05e{-}07$	2.20	$2.15e{-}09$	2.88	$1.72e{-}12$	1.52

Table 8.8: Test example 6: L_2 convergence for constant stepsizes

8.3 Adaptive Discretization in Time

Concerning the numerical results with non–constant time step sizes h, we shall distinguish the cases of a full and of an adaptive spatial discretization.

8.3.1 Full Spatial Discretization

In order to check the performance of the stepsize controller, we collect some numerical results for the one–dimensional cases 1 to 6. As in Section 8.2, we use a quadratic spline wavelet basis ($m = \tilde{m} = 3$) on the interval from [126], and the spatial discretization is done using all wavelets up to the maximal refinement level $j_{max} = 15$. For tolerances TOL from 2^{-6} to 2^{-25}, an adaptive time discretization was applied to the test problems, where we solved each elliptic subproblem by a single Galerkin projection onto the space $V_{j_{max}+1}$ of active wavelets.

Since for an adaptive discretization in time with the considered one–step methods, there is no general proof of the attainable convergence order available, we shall only report the numerical results in graphical form in the following. In the case of equidistant stepsizes, the number of time steps N is proportional to h^{-1}. Consequently, using an adaptive stepsize controller, we hope that the L_2 accuracy

$$\max_n \|u(t_n) - u^{(n)}\|_{L_2(\Omega)} \tag{8.3.1}$$

behaves like a constant times $N^{-p_{num}}$, where p_{num} may coincide or not with the classical order p of the corresponding ROW–method. For each of the test examples, we give a plot of the required number of time steps against the resulting L_2 accuracy. Secondly, we compare this L_2 error with the CPU time needed to compute the approximations. Due to the fixed spatial discretization, the computational work per time step is uniform, so that both curves should only differ by a vertical shift. Hence the slopes are only measured in the respective left diagram.

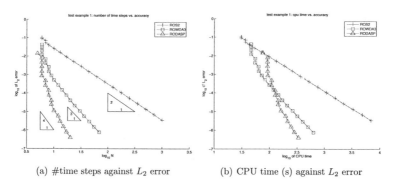

(a) #time steps against L_2 error   (b) CPU time (s) against L_2 error

Figure 8.9: Test example 1: L_2 convergence for a time–adaptive discretization.

The results for example 1 can be found in Figure 8.9. It is clearly visible that all integrators show the respective classical integer convergence orders $p \in \{2, 3, 4\}$, similar to the case of constant stepsizes.

In the second test example, recall that we observed a severe order reduction for equidistant stepsizes due to the Hölder singularity of $u : [0, T] \to L_2(\Omega)$ at the origin. For an adaptive discretization in time, this is no longer the case. Instead, we can measure numerical rates of convergence p_{num} of approximately 1.75, 2.75 and 3.75 for the integrators under consideration, see Figure 8.10.

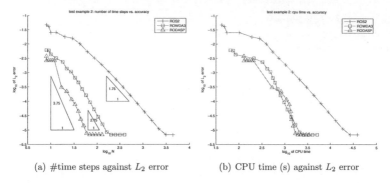

(a) #time steps against L_2 error　　　(b) CPU time (s) against L_2 error

Figure 8.10: Test example 2: L_2 convergence for a time–adaptive discretization.

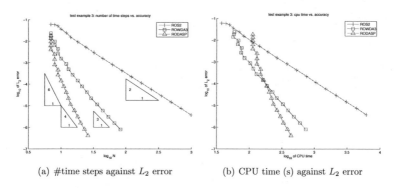

(a) #time steps against L_2 error　　　(b) CPU time (s) against L_2 error

Figure 8.11: Test example 3: L_2 convergence for a time–adaptive discretization.

In Figure 8.11, we report the convergence results for test example 3. As in the case of constant stepsizes, we can observe the classical convergence orders $p \in \{2, 3, 4\}$ also for an adaptive discretization in time. Surprisingly, the fourth–order scheme RODASP even shows a higher numerical convergence rate p_{num} of approximately 6.

In test example 4, where the spatial regularity of f and u was very low, we have observed a severely limited convergence rate in the case of equidistant stepsizes, see Section 8.2. For an adaptive choice of h, the temporal convergence rate considerably improves and we can again observe at least the classical rates, see Figure 8.12. Again the scheme RODASP performs very well, with a rate $p_{num} \approx 5$.

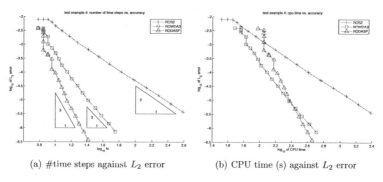

(a) #time steps against L_2 error   (b) CPU time (s) against L_2 error

Figure 8.12: Test example 4: L_2 convergence for a time–adaptive discretization.

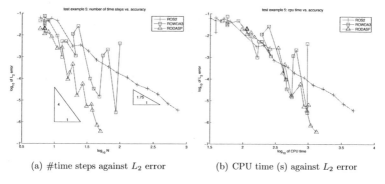

(a) #time steps against L_2 error   (b) CPU time (s) against L_2 error

Figure 8.13: Test example 5: L_2 convergence for a time–adaptive discretization.

The test example 5 had been chosen in such a way that f has a discontinuity at $t = \frac{1}{2}$ and $f(t)$ is discontinuous in space for $t < \frac{1}{2}$, leading to a suboptimal convergence behavior of the considered ROW–methods. The corresponding numerical results are presented in Figure 8.13. For the schemes ROS2 and ROWDA3, we can observe a rate $p_{num} \approx 1.75$ although it should be noted that the third–order method ROWDA3 behaves rather irregular in this test example. For RODASP, the classical rate 4 is attained. Summing up, the convergence rates for an adaptive discretization in time clearly outperform the results from the case of constant stepsizes. It becomes clear that example 5 is one of the problems where temporal adaptivity really pays off. This becomes even more apparent when comparing the associated computational work for adaptive and nonadaptive time integration, see Figure 8.21.

Finally, we present the results for test example 6 in Figure 8.14. It is visible that the schemes ROS2 and RODASP show at least the classical convergence rates $p \in \{2, 4\}$, respectively. The second–order scheme ROS2 performs even better, with $p_{num} \approx 2.25$. In analogy to the results for constant stepsizes, however, the scheme ROWDA3 did not yield satisfactory results. We observed a numerical rate strictly

less than 2. In contrast to ROWDA3, the scheme ROS3Pw worked perfectly also in
the case of adaptive stepsizes, with $p_{num} \approx 3$.

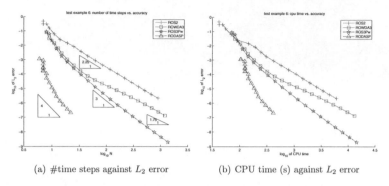

(a) #time steps against L_2 error   (b) CPU time (s) against L_2 error

Figure 8.14: Test example 6: L_2 convergence for a time–adaptive discretization.

8.3.2   Fully Adaptive Discretization

We now turn to the numerical results for a fully adaptive discretization in time and
space. Both to the one– and to the two–dimensional test examples, we have applied
an adaptive time discretization for tolerances TOL from 2^{-6} to 2^{-20}.

1D Results

On the interval, we choose again the quadratic spline wavelet basis ($m = \tilde{m} = 3$)
on the interval from [126]. The spatial discretization is done adaptively, where we
restrict the wavelet basis to all wavelets with $|\lambda| \leq j_{max} = 12$. Of course this
approach somehow contradicts a fully adaptive setting, but for the moment this was
the most feasible approach concerning the software implementation. Future versions
of the code should make use of tree structured wavelet index sets where the artificial
limitation can be avoided.

Figure 8.15 shows the convergence and complexity results for test example 1.
Similar to the case of a full spatial discretization, the number of time steps N for
a given accuracy ε behaves like $N \approx \varepsilon^{-1/p}$, for $p \in \{2, 3, 4\}$, respectively. Concerning the asymptotic behaviour of the computational work, we observe slightly better
rates of approximately 2.25, 3.25 and 5.25. This is due to the fact that for homogeneous problems, the computational work for time steps t_n close to 0 is substantially
higher than for later time steps due to the parabolic smoothing, which results in a
decreasing complexity per step as t_n tends to T. Consequently, an adaptive spatial
discretization really pays off compared to a uniform one in the case of homogeneous
problems.

In the computations for test example 2, see Figure 8.16, we can see that the
schemes ROS2 and ROWDA3 exhibit a slightly worse convergence behavior compared with a full spatial discretization. In both cases, the measured rates are about

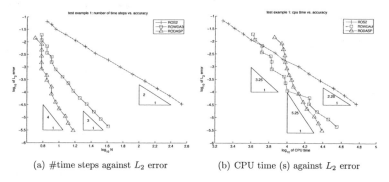

(a) #time steps against L_2 error   (b) CPU time (s) against L_2 error

Figure 8.15: Test example 1: L_2 convergence for a fully adaptive discretization.

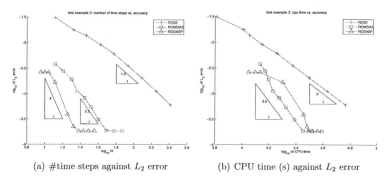

(a) #time steps against L_2 error   (b) CPU time (s) against L_2 error

Figure 8.16: Test example 2: L_2 convergence for a fully adaptive discretization.

0.25 smaller for a fully adaptive approximation. This may be explained by the additional truncation error when using only wavelets with a level $|\lambda| \leq 12$ in the spatially adaptive experiments compared with $|\lambda| \leq 15$ for the spatially nonadaptive case. The integrator RODASP shows a fourth order convergence which is better than in the spatially non–adaptive experiments. Concerning the computational work, we again observe for each scheme that the asymptotic rate of the CPU time is better than the corresponding rate of timesteps N. The chosen integrators show complexity rates of 2, 3.5 and approximately 4, respectively, outperforming the complexity behavior in the case of a full spatial discretization. For example 2, we hence note that the additional spatial adaptivity indeed helps, although this is not yet visible when only considering the number of time steps.

For test example 3, we do not expect an improvement of the convergence and complexity behavior compared to a non–adaptive setting since the solution u is uniformly smooth both in time and space. The numerical results are given in Figure 8.17. In fact, we can observe rates of 2 for ROS2 and 3 for ROWDA3, as in the non–adaptive discretizations. The scheme RODASP shows a convergence of order

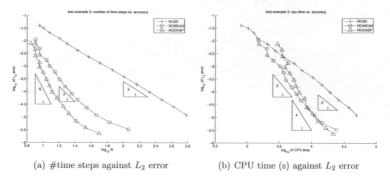

(a) #time steps against L_2 error   (b) CPU time (s) against L_2 error

Figure 8.17: Test example 3: L_2 convergence for a fully adaptive discretization.

6 before the spatial discretization error begins to dominate the overall scheme. The complexity plot clearly shows that the computational work behaves like $\varepsilon^{-1/p}$, where $p \in \{2, 3, 4\}$ are the classical convergence orders of the considered ROW–methods.

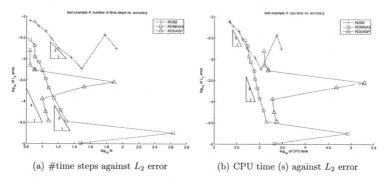

(a) #time steps against L_2 error   (b) CPU time (s) against L_2 error

Figure 8.18: Test example 4: L_2 convergence for a fully adaptive discretization.

For example 4, the convergence and complexity plots look a bit irregular, see Figure 8.18. For low and moderate tolerances, the schemes ROS2 and ROWDA3 exhibit a convergence order 2 and 3, respectively, similar to the spatially non–adaptive discretization. For small tolerances, we still have convergence but the measured accuracies do not allow for a reliable order estimation. Also the scheme RODASP shows a fourth order convergence for low tolerances, with an oscillating behavior for smaller values of TOL as for the lower–order schemes. The complexity curves look very similar, up to a vertical shift.

Concerning the results for example 5, the schemes ROS2 and RODASP have a convergence behavior that is at least as good as in the non–adaptive case. In Figure 8.19, we can observe a numerical rate of approximately 1.75 for ROS2 and even 6 for RODASP. The integrator ROWDA3 behaves more irregular. Here, for

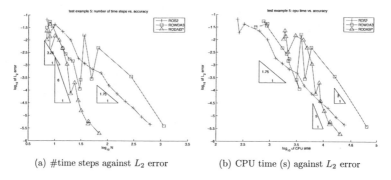

(a) #time steps against L_2 error   (b) CPU time (s) against L_2 error

Figure 8.19: Test example 5: L_2 convergence for a fully adaptive discretization.

low tolerances, a rate $p_{num} \approx 3$ is obtained, whereas the results become oscillating for lower values of TOL. It seems that the embedded error estimator in ROWDA3 is not as reliable as the ones of the other ROW–methods, since we used the identical stepsize controller for all integrations. The computational work shows a comparable asymptotic behaviour as TOL decreases.

Due to the good convergence behavior of adaptive methods in test example 5, we have also compared the absolute computational complexity of the three presented methods of different degrees of adaptivity in Figure 8.21. We clearly observe the different complexity rates of nonadaptive methods on the one hand and temporally adaptive schemes on the other hand. The breakeven point between temporally adaptive and nonadaptive schemes is reached already at a moderate tolerance, in particular for ROS2 and RODASP. However, due to the fact that the solution of the full 1D Galerkin system is comparatively cheap, spatial adaptivity does not really pay off yet in terms of CPU time.

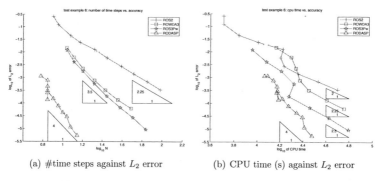

(a) #time steps against L_2 error   (b) CPU time (s) against L_2 error

Figure 8.20: Test example 6: L_2 convergence for a fully adaptive discretization.

Finally, the results for the test example 6 are shown in Figure 8.20. The number of time steps N shows the same asymptotic behavior as in the case of a full spatial discretization. For the schemes ROS2 and RODASP, we observe a rate of 2.25 and 4, respectively. Moreover, the third–order scheme ROWDA3 does again not perform well, we measure an inferior rate of approximately 1.75. This is not the case for the integrator ROS3Pw which we already considered as a benchmark scheme in the case of equidistant stepsizes. Here we observe a high convergence rate of approximately 3.5. The computational complexity of ROS2 and RODASP attains the integer rates 2 and 4, respectively, whereas the third order schemes perform at an approximate slope 2.5. Compared to a full spatial discretization, the computational cost for the elliptic subproblems in the fully adaptive scheme seems to grow as t_n tends to T, which is presumably due to error propagation.

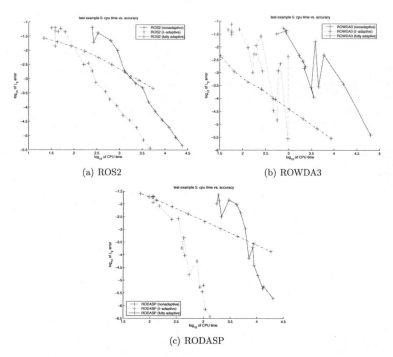

(a) ROS2   (b) ROWDA3

(c) RODASP

Figure 8.21: Test example 5: computational work diagram for nonadaptive (*dash–dotted lines*), time–adaptive (*dotted lines*) and fully adaptive (*solid lines*) discretizations.

2D Results

Finally, we present some numerical examples for a fully adaptive discretization of the test problems 7 and 8 on the L–shaped domain $\Omega = (-1,1)^2 \setminus [0,1)^2$. As a wavelet basis, we choose a linear spline composite basis ($m = \tilde{m} = 2$), where the internal 1D wavelet basis is taken from [62] with the stabilization of the wavelets as proposed in [9]. Similar to the one–dimensional tests, the spatial discretization uses a subset of the overall wavelet basis up to a maximal refinement level of $j_{\max} = 5$, i.e., the spatial approximations are contained in the multiresolution space V_6. This restriction will clearly have an effect on the adaptive solutions of the stage equations, but due to the large number of time steps (approximately 100) for small tolerances TOL, we decided to constrain the elliptic solver in such a way to keep the runtime of the code under a reasonable size. For a nonadaptive Galerkin solver, as used in the one–dimensional examples or in the experiments of [103], it is possible to choose a higher maximal refinement level since the computational work per time step stays relatively small as long as the stiffness matrices are precomputed.

As already stated above, we shall only use second and third–order ROW-methods for the numerical experiments. This is due to the fact that the high number of 6 stages for RODASP leads to a considerable computational work.

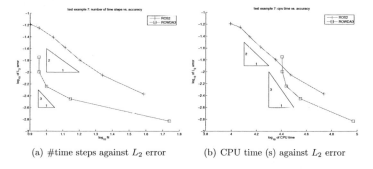

(a) #time steps against L_2 error (b) CPU time (s) against L_2 error

Figure 8.22: Test example 7: L_2 convergence for a fully adaptive discretization.

In Figure 8.22, we report the numerical results for the test example 7. Here the driving term f and the solution u are temporally and spatially smooth. For the second–order scheme ROS2, the number of time steps for the adaptive discretization behaves like $\varepsilon^{-1/2}$, as it would be the case for equidistant time steps. The scheme ROWDA3, however, does not yield satisfactory results. For low tolerances, the number of time steps increases like $\varepsilon^{-1/3}$, but the spatial discretization errors become dominant quickly as the tolerances decrease. The complexity plots show the same behavior. We observe a rate of 2 for ROS2, whereas ROWDA3 exhibits a complexity rate of approximately 3 as long as the tolerances are low.

The results for example 8 are given in Figure 8.23. Here the temporal singularity at $t = 0$ and the spatial corner singularity affect the numerical convergence rate of ROS2. We observe a number of time steps N increasing like $\varepsilon^{-1/1.25}$. In comparison, similar to the example 7, the ROWDA3 scheme shows a rate of 3 first, but the overall

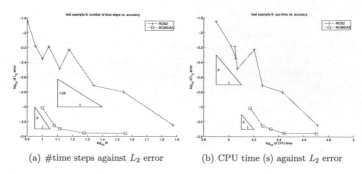

(a) #time steps against L_2 error   (b) CPU time (s) against L_2 error

Figure 8.23: Test example 8: L_2 convergence for a fully adaptive discretization.

convergence is quickly deteriorated by the spatial discretization errors. Concerning the computational complexity, it is clearly visible that the elliptic subproblems for the integrator ROS2 become less expensive as t_n reaches T, which leads in a complexity rate of approximately 2. Obviously, spatial adaptivity is helpful for problems of type 8, see also the findings in [83]. Finally, the scheme ROWDA3 quickly reaches the lowest possible accuracy at an initial rate of approximately 3. Moreover, the CPU times behave like $\varepsilon^{-1/2}$ for small tolerances.

Concluding Remarks

In this thesis, we have studied several extensions to the current state of research in adaptive wavelet methods. For a systematic classification of the results, we shall revisit the guiding tasks (T1)–(T3) that were formulated in the introduction.

First of all, in order to circumvent the potentially complicated construction of numerically stable wavelet bases on a bounded polygonal domain $\Omega \subset \mathbb{R}^d$, we were concerned with the alternative concept of wavelet frames, according to task (T1). For the discretization of elliptic operator equations with frames, it has turned out that the class of wavelet Gelfand frames is a particular convenient one, see Section 2.1. Gelfand frames are numerically stable systems in L_2 that allow for the characterization of Sobolev and Besov spaces just by an appropriate rescaling, completely similar to the case for wavelet Riesz bases. The latter are, in fact, a subclass of Gelfand frames. After the theoretical specification of this convenient frame concept, we were able to show that there are indeed practical constructions of Gelfand frames on domains with a nontrivial geometry. Inspired by [133], we have studied overlapping domain decompositions of Ω into subpatches that are parametric images of the cube. This approach immediately induces aggregated wavelet frames, taking the union of appropriately lifted reference bases. Although it has been straightforward to verify that the overall system is an L_2 frame, the proof of the Gelfand frame property and hence also that of the characterization of function spaces required the application and nontrivial extension of the localization theory of frames, see Section 2.2. By their construction, aggregated wavelet frames retain the locality, regularity and cancellation properties of the reference basis, which can be exploited in numerical applications. Besides their analytic properties, it is most important that the construction of aggregated wavelet frames is simple, having a positive influence on the corresponding computer code. Whereas the numerical implementation of, say, the composite wavelet bases from [62] is a rather painful task, aggregated wavelet frames are available as soon as the appropriate parametric mappings and a well–conditioned reference basis on the cube is implemented. As a consequence, we can say that problem (T1) has been solved to the full extent.

After the construction of suitable wavelet frames on a polygonal domain, we have discussed their application to elliptic operator equations. First of all, in Section 5.1, the equivalent reformulation of the original operator equation in wavelet Gelfand frame coordinates was given. Unlike the case of Riesz bases, here we are confronted with a singular system matrix, which results from the redundancy of the underlying frame. For the design of convergent adaptive frame methods, the singularity of the system matrix is not an issue. As the most striking adaptive method, we were able to specify an approximative descent iteration of Richardson type that is guaranteed to converge for symmetric stationary elliptic problems, see Section 5.2, The kernel of the biinfinite stiffness matrix indeed comes into play in the complexity analysis of the overall scheme. Due to the fact that the iterands in the adaptive Richardson algorithm are no longer contained in the respective Krylov spaces, kernel components may accumulate during the iteration. However, under a technical assumption which can be proven to hold in special cases, optimality of the adaptive frame algorithm can be established. Finally, in Section 5.3, we have tested the adaptive Richardson iteration in several numerical examples in one and two space dimensions, validating the claimed convergence and complexity properties of the overall algorithm. Of course, as already stated in Section 5.3, the quantitative results of adaptive frame schemes may be improved by looking at alternative variants like approximate steepest descent schemes, see [52]. Summing up, also the aspects of task (T2) have been addressed completely. As a consequence, instead of constructing stable wavelet bases the potential advantages of which are wasted by their complicated implementation, we therefore recommend the use of suitable wavelet Gelfand frames as one possible alternative.

As another major topic, we have addressed the application of wavelet methods to the adaptive numerical solution of linear parabolic problems. According to task (T3), we have studied a well–established two–step strategy which consists in a semidiscretization in time and a successive spatial discretization. In contrast to the prototypical schemes from [107], we have employed wavelet methods for the spatial approximation. The analytic properties of wavelet bases such as the characterization of function spaces can be utilized in the numerical algorithm. By an appropriate coupling of a linearly implicit time integrator with well–known adaptive wavelet algorithms for the elliptic subproblems, we obtained a fully adaptive numerical scheme. An adaptive increment algorithm has been specified in Section 7.2 and its convergence and complexity properties have been analyzed. The convergence of the global algorithm, though, relies on the convergence of the stepsize controller, since spatial errors are interpreted as additional temporal errors in the iteration. Finally, we have presented several numerical examples in one and two space dimensions that support the theoretical analysis. In all, we regard the task (T3) as solved, though quantitative improvements of the current implementation are still necessary to fully exploit the advantages of the proposed adaptive method.

In any of the three discussed topics, there are future perspectives of the presented results. Firstly, the theoretical properties of the considered frame construction are not fully settled at the moment. As an example, the characterization of the full Besov scale by aggregated wavelet frame coefficients is an open problem, compared to the case of wavelet bases. Current investigations show that there is space for improvement [50]. Moreover, the proof of the Gelfand frame property may be drastically simplified by using more a priori knowledge on the reference wavelet basis and on the particular parametrization of the domain.

Secondly, the consideration of an overlapping domain decomposition may be further exploited by using alternative approximation methods than biinfinite linear iterative schemes, as already mentioned in the introduction. An apparent perspective is the development of overlapping domain decomposition algorithms using wavelet frames, similar to methods known from finite element methods. First numerical tests with additive and multiplicative Schwarz iterations in a current thesis [141] have shown that methods of this type are superior to all known frame algorithms, at least quantitatively. Of course, it would be interesting to compare this particular kind of frame schemes also with non–overlapping domain decomposition techniques. Both alternatives have the important feature that they are highly parallelizable, which is of ultimate importance when considering realistic problems from technical applications. Furthermore, of course, the range of problems which the considered frame schemes apply to may be extended. By their similarity to wavelet bases, aggregated wavelet frames should allow also for the treatment of nonsymmetric and indefinite problems, in the spirit of [34, 47, 81]. Moreover, a particularly interesting open field is the extension of frame methods to nonlinear problems.

Concerning the application of wavelet methods to parabolic problems, there are several possibilities to enhance both the global method but also its building blocks. To begin with the latter, a natural question is whether it is possible to replace the internal wavelet methods by a frame algorithm, possibly also one of the aforementioned domain decomposition schemes. From the theoretical point of view, this does not seem to be a problem. We expect a considerable quantitative improvement by the use of frame methods due to the simpler numerical implementation. Furthermore, the class of considered time integrators may be exchanged by alternative ones. We have mentioned in Remark 7.1 that, in view of the matrix compression properties of wavelet bases and frames, also W–methods that use inexact Jacobians may be of interest. Alternatively, it is also an interesting prospect to substitute the Rosenbrock methods by an appropriate multistep integrator [84], circumventing the problem of order reduction. Finally, it should be noted that we have explicitly considered a class of algorithms that permits the generalization towards nonlinear parabolic problems. For a corresponding numerical realization, one essentially needs algorithms to evaluate nonlinear expressions of wavelet expansions. Potential studies into the direction of nonlinear parabolic problems may rely here on the adaptive evaluation methods presented in [6, 36].

List of Figures

List of Tables

Bibliography

[1] R. A. Adams, *Sobolev spaces*, Academic Press, New York, 1975.

[2] H. Amann, *Linear and quasilinear parabolic problems. Vol. 1: Abstract linear theory*, Monographs in Mathematics, vol. 189, Birkhäuser, Basel, 1995.

[3] L. Angermann and J. Rang, *New Rosenbrock W-methods of order 3 for partial differential algebraic equations of index 1*, BIT **45** (2005), no. 4, 761–787.

[4] E. Arendt, *Semigroups and evolution equations: Functional calculus, regularity and kernel estimates*, Evolutionary Equations (C.M. Dafermos et al., ed.), vol. 1, Elsevier/North–Holland, 2004, pp. 1–85.

[5] A. Averbuch, A. Cohen, and M. Israeli, *A stable and accurate explicit scheme for parabolic evolution equations*, preprint, 1998.

[6] A. Barinka, *Fast computation tools for adaptive wavelet schemes*, Ph.D. thesis, RWTH Aachen, 2005.

[7] A. Barinka, T. Barsch, P. Charton, A. Cohen, S. Dahlke, W. Dahmen, and K. Urban, *Adaptive wavelet schemes for elliptic problems: Implementation and numerical experiments*, SIAM J. Sci. Comput. **23** (2001), no. 3, 910–939.

[8] A. Barinka, T. Barsch, S. Dahlke, and M. Konik, *Some remarks on quadrature formulas for refinable weight functions*, Zeitschrift für Angewandte Mathematik und Mechanik **81** (2001), 839–855.

[9] T. Barsch, *Adaptive Multiskalenverfahren für elliptische partielle Differential-gleichungen – Realisierung, Umsetzung und numerische Ergebnisse*, Ph.D. thesis, RWTH Aachen, 2001.

[10] C. Bennett and R. Sharpley, *Interpolation of operators*, Pure and Applied Mathematics, vol. 129, Academic Press, 1988.

[11] J. Bergh and J. Löfström, *Interpolation Spaces*, Grundlehren der mathematischen Wissenschaften, vol. 223, Springer–Verlag, 1976.

[12] T. Betcke and L.N. Trefethen, *Reviving the method of particular solutions*, SIAM Rev. **47** (2005), no. 3, 469–491.

[13] G. Beylkin, *Wavelets and fast numerical algorithms*, Different perspectives on wavelets. American Mathematical Society short course on wavelets and applications, held in San Antonio, TX (USA), January 11-12, 1993 (Providence, RI) (I. Daubechies, ed.), American Mathematical Society, 1993, pp. 89–117.

[14] P. Binev, W. Dahmen, and R. DeVore, *Adaptive finite element methods with convergence rates*, Numer. Math. **97** (2004), no. 2, 219–268.

[15] K. Bittner, *Biorthogonal spline wavelets on the interval*, Wavelets and splines: Athens 2005, Mod. Methods Math., Nashboro Press, Brentwood, TN, 2006, pp. 93–104.

[16] J.G. Blom, W. Hundsdorfer, E.J. Spee, and J.G. Verwer, *A second-order Rosenbrock method applied to photochemical dispersion problems*, SIAM J. Sci. Comput. **20** (1999), no. 4, 1456–1480.

[17] F. Bornemann, *Adaptive multilevel discretization in time and space for parabolic partial differential equations*, Bericht TR 89-07, Konrad-Zuse-Zentrum Berlin, 1989.

[18] ——, *An adaptive multilevel approach to parabolic equations I. General theory and 1D implementations*, Impact Comput. Sci. Engrg. **2** (1990), 279–317.

[19] ——, *An adaptive multilevel approach for parabolic equations in two space dimensions*, Bericht TR 91-07, Konrad-Zuse-Zentrum Berlin, 1991.

[20] ——, *An adaptive multilevel approach to parabolic equations II. Variable-order time discretization based on a multiplicative error correction*, Impact Comput. Sci. Engrg. **3** (1991), 93–122.

[21] F. Bornemann and P. Deuflhard, *Numerische Mathematik II*, Walter de Gruyter, Berlin–New York, 2002.

[22] L. Borup, R. Gribonval, and M. Nielsen, *Bi-framelet systems with few vanishing moments characterize Besov spaces*, Appl. Comput. Harmon. Anal. **17** (2004), 3–28.

[23] ——, *Nonlinear approximation with bi-framelets*, Proceedings of the 11th international conference, Gatlinburg, TN, USA, May 18–22, 2004 (Brentwood, TN) (C.K. Chui et al., ed.), 2005.

[24] C. Canuto, A. Tabacco, and K. Urban, *Numerical solution of elliptic problems by the wavelet element method*, ENUMATH 97. Proceedings of the 2nd European conference on numerical mathematics and advanced applications held in Heidelberg, Germany, September 28-October 3, 1997. Including a selection of papers from the 1st conference (ENUMATH 95) held in Paris, France, September 1995 (Singapore) (H.G. Bock et al., eds.), World Scientific, 1998, pp. 17–37.

[25] ——, *The wavelet element method, part I: Construction and analysis*, Appl. Comput. Harmon. Anal. **6** (1999), 1–52.

[26] _____, *The wavelet element method, part II: Realization and additional features in 2D and 3D*, Appl. Comput. Harmon. Anal. **8** (2000), 123–165.

[27] C. Canuto and K. Urban, *Adaptive optimization in convex Banach spaces*, SIAM J. Numer. Anal. **42** (2005), no. 5, 2043–2075.

[28] J. M. Carnicer, W. Dahmen, and J. M. Peña, *Local decomposition of refinable spaces and wavelets*, Appl. Comput. Harmon. Anal. **3** (1996), 127–153.

[29] O. Christensen, *An Introduction to Frames and Riesz Bases*, Birkhäuser, 2003.

[30] C. K. Chui and E. Quak, *Wavelets on a bounded interval*, Numerical Methods of Approximation Theory (Dietrich Braess and Larry L. Schumaker, eds.), International Series of Numerical Mathematics, vol. 9, Birkhäuser, 1992, pp. 53–75.

[31] A. Cohen, *Wavelet methods in numerical analysis*, Handbook of Numerical Analysis (P.G. Ciarlet and J.L. Lions, eds.), vol. VII, North–Holland, Amsterdam, 2000, pp. 417–711.

[32] _____, *Numerical analysis of wavelet methods*, Studies in Mathematics and its Applications, vol. 32, North–Holland, Amsterdam, 2003.

[33] A. Cohen, W. Dahmen, and R. DeVore, *Adaptive wavelet methods for elliptic operator equations — Convergence rates*, Math. Comput. **70** (2001), no. 233, 27–75.

[34] _____, *Adaptive wavelet methods II: Beyond the elliptic case*, Found. Comput. Math. **2** (2002), no. 3, 203–245.

[35] _____, *Adaptive wavelet schemes for nonlinear variational problems*, SIAM J. Numer. Anal. **41** (2003), no. 5, 1785–1823.

[36] _____, *Sparse evaluation of compositions of functions using multiscale expansions*, SIAM J. Math. Anal. **35** (2003), no. 2, 279–303.

[37] A. Cohen, I. Daubechies, and J.-C. Feauveau, *Biorthogonal bases of compactly supported wavelets*, Commun. Pure Appl. Math. **45** (1992), 485–560.

[38] A. Cohen, I. Daubechies, and P. Vial, *Wavelets on the interval and fast wavelet transforms*, Appl. Comput. Harmon. Anal. **1** (1993), 54–81.

[39] A. Cohen and R. Masson, *Wavelet methods for second-order elliptic problems, preconditioning, and adaptivity*, SIAM J. Sci. Comput. **21** (1999), no. 3, 1006–1026.

[40] _____, *Wavelet adaptive method for second order elliptic problems: Boundary conditions and domain decomposition*, Numer. Math. **86** (2000), no. 2, 193–238.

[41] E. Cordero and K. Gröchenig, *Localization of frames II*, Appl. Comput. Harmon. Anal. **17** (2004), no. 1, 29–47.

[42] T. H. Cormen, C. E. Leiserson, R. L. Rivest, and C. Stein, *Introduction to algorithms*, second ed., MIT Press and McGraw–Hill, 2001.

[43] S. Dahlke, *Wavelets: Construction principles and applications to the numerical treatment of operator equations*, Shaker, Aachen, 1997.

[44] _____, *Besov regularity for elliptic boundary value problems on polygonal domains*, Applied Mathematics Letters **12** (1999), 31–38.

[45] S. Dahlke, W. Dahmen, and R. DeVore, *Nonlinear approximation and adaptive techniques for solving elliptic operator equations*, Multiscale Wavelet Methods for Partial Differential Equations (W. Dahmen, A. Kurdila, and P. Oswald, eds.), Academic Press, San Diego, 1997, pp. 237–283.

[46] S. Dahlke, W. Dahmen, R. Hochmuth, and R. Schneider, *Stable multiscale bases and local error estimation for elliptic problems*, Appl. Numer. Math. **23** (1997), 21–48.

[47] S. Dahlke, W. Dahmen, and K. Urban, *Adaptive wavelet methods for saddle point problems — Optimal convergence rates*, SIAM J. Numer. Anal. **40** (2002), no. 4, 1230–1262.

[48] S. Dahlke and R. DeVore, *Besov regularity for elliptic boundary value problems*, Commun. Partial Differ. Equations **22** (1997), no. 1&2, 1–16.

[49] S. Dahlke, M. Fornasier, and K.-H. Gröchenig, *Optimal adaptive computations in the Jaffard algebra and localized frames*, Preprint 146, DFG Schwerpunktprogramm 1114 *Mathematische Methoden der Zeitreihenanalyse und digitalen Bildverarbeitung*, 2006.

[50] S. Dahlke, M. Fornasier, M. Primbs, T. Raasch, and M. Werner, *Nonlinear approximation with Gelfand frames*, in preparation, 2007.

[51] S. Dahlke, M. Fornasier, and T. Raasch, *Adaptive frame methods for elliptic operator equations*, Adv. Comput. Math. (2006), published online, DOI 10.1007/s10444-005-7501-6.

[52] S. Dahlke, M. Fornasier, T. Raasch, R. Stevenson, and M. Werner, *Adaptive frame methods for elliptic operator equations: The steepest descent approach*, Bericht 2006-2, Fachbereich Mathematik und Informatik, Philipps-Universität Marburg, 2006, To appear in *IMA J. Numer. Anal.*

[53] W. Dahmen, *Multiscale analysis, approximation, and interpolation spaces*, Approximation Theory VIII (C. K. Chui, ed.), vol. 2, 1995, pp. 47–88.

[54] _____, *Stability of multiscale transformations*, J. Fourier Anal. Appl. **4** (1996), 341–362.

[55] _____, *Wavelet and multiscale methods for operator equations*, Acta Numerica **6** (1997), 55–228.

[56] W. Dahmen, B. Han, R. Q. Jia, and A. Kunoth, *Biorthogonal multiwavelets on the interval: Cubic Hermite splines*, Constr. Approx. **16** (2000), no. 2, 221–259.

[57] W. Dahmen and A. Kunoth, *Multilevel preconditioning*, Numer. Math. **63** (1992), no. 3, 315–344.

[58] W. Dahmen, A. Kunoth, and K. Urban, *Wavelets in numerical analysis and their quantitative properties*, Surface fitting and multiresolution methods (A. Le Méhauté, C. Rabut, and L. Schumaker, eds.), vol. 2, Vanderbilt University Press, 1997, pp. 93–130.

[59] _____, *Biorthogonal spline–wavelets on the interval — Stability and moment conditions*, Appl. Comput. Harmon. Anal. **6** (1999), 132–196.

[60] W. Dahmen, S. Prössdorf, and R. Schneider, *Multiscale methods for pseudodifferential operators on smooth manifolds*, Proceedings of the International Conference on Wavelets: Theory, Algorithms and Applications (C.K. Chui, L. Montefusco, and L. Puccio, eds.), Academic Press, 1994, pp. 385–424.

[61] W. Dahmen and R. Schneider, *Wavelets with complementary boundary conditions — Function spaces on the cube*, Result. Math. **34** (1998), no. 3–4, 255–293.

[62] _____, *Composite wavelet bases for operator equations*, Math. Comput. **68** (1999), 1533–1567.

[63] _____, *Wavelets on manifolds I. Construction and domain decomposition*, SIAM J. Math. Anal. **31** (1999), 184–230.

[64] W. Dahmen, J. Vorloeper, and K. Urban, *Adaptive wavelet methods — basic concepts and applications to the Stokes problem*, Proceedings of the International Conference of Computational Harmonic Analysis (D.-X. Zhou, ed.), World Scientific, 2002, pp. 39–80.

[65] I. Daubechies, *Orthonormal bases of compactly supported wavelets*, Commun. Pure Appl. Math. **41** (1988), no. 7, 909–996.

[66] _____, *Ten lectures on wavelets*, CBMS–NSF Regional Conference Series in Applied Math., vol. 61, SIAM, Philadelphia, 1992.

[67] C. de Boor, *A practical guide to splines*, revised ed., Applied Mathematical Sciences, vol. 27, Springer, New York, 2001.

[68] P. Deuflhard and A. Hohmann, *Numerische Mathematik I*, Walter de Gruyter, Berlin–New York, 2002.

[69] R. DeVore, *Nonlinear approximation*, Acta Numerica **7** (1998), 51–150.

[70] R. DeVore and G.G. Lorentz, *Constructive approximation*, Springer, Berlin, 1998.

[71] George C. Donovan, Jeffrey S. Geronimo, and Douglas P. Hardin, *Intertwining multiresolution analyses and the construction of piecewise polynomial wavelets*, SIAM J. Appl. Math. **27** (1996), no. 6, 1791–1815.

[72] ———, *Orthogonal polynomials and the construction of piecewise polynomial smooth wavelets*, SIAM J. Appl. Math. **30** (1999), no. 5, 1029–1056.

[73] R. J. Duffin and A. C. Schaeffer, *A class of nonharmonic Fourier series*, Trans. Amer. Math. Soc. **72** (1952), 341–366.

[74] N. Dunford and J.T. Schwartz, *Linear operators. I. General theory*, Wiley, New York, 1958.

[75] M. Ehler, *On the construction of compactly supported nonseparable bi-frames and their best n–term approximation properties*, Dissertation, Philipps–Universität Marburg, in preparation.

[76] W. Fischer and I. Lieb, *Function theory. Complex analysis in one variable*, Vieweg Studium: Aufbaukurs Mathematik, Vieweg, Wiesbaden, 2005.

[77] M. Fornasier and K. Gröchenig, *Intrinsic localization of frames*, Constr. Approx. **22** (2005), no. 3, 395–415.

[78] L. Fox, P. Henrici, and C. Moler, *Approximations and bounds for eigenvalues of elliptic operators*, SIAM J. Numer. Anal. **4** (1967), no. 1, 89–102.

[79] M. Frazier and B. Jawerth, *A discrete transform and decompositions of distribution spaces*, J. Funct. Anal. **93** (1990), no. 1, 34–170.

[80] D. Fujiwara, *Concrete characterization of the domains of fractional powers of some elliptic differential operators of the second order*, Proc. Japan Acad. **43** (1967), no. 2, 82–86.

[81] T. Gantumur, *An optimal adaptive wavelet method for nonsymmetric and indefinite elliptic problems*, Tech. Report 1343, Utrecht University, January 2006, Submitted.

[82] T. Gantumur, H. Harbrecht, and R. Stevenson, *An optimal adaptive wavelet method without coarsening of the iterands*, Tech. Report 1325, Utrecht University, March 2005, To appear in *Math. Comp.*

[83] K. Gerdes, D. Schötzau, C. Schwab, and T. Werder, *hp–discontinuous Galerkin time stepping for parabolic problems*, Comput. Methods Appl. Mech. Eng. **190** (2001), no. 49–50, 6685–6708.

[84] A. Gerisch, J. Lang, H. Podhaisky, and R. Weiner, *High–order finite element–linear implicit two–step peer methods for time–dependent PDEs*, Tech. Report 13, MLU Halle–Wittenberg, 2006.

[85] P. Grisvard, *Singularities in boundary value problems*, Research Notes in Applied Mathematics, Springer, 1992.

[86] K.-H. Gröchenig, *Describing functions: Atomic decompositions versus frames,* Monatsh. Math. **112** (1991), no. 1, 1–42.

[87] ———, *Localization of frames,* GROUP 24: Physical and Mathematical Aspects of Symmetries (Bristol) (J.-P. Gazeau, R. Kerner, J.-P. Antoine, S. Metens, and J.-Y. Thibon, eds.), IOP Publishing, 2003, to appear.

[88] ———, *Localized frames are finite unions of Riesz sequences,* Adv. Comput. Math. **18** (2003), no. 2-4, 149–157.

[89] ———, *Localization of frames, Banach frames, and the invertibility of the frame operator,* J. Fourier Anal. Appl. **10** (2004), no. 2, 105–132.

[90] K. Gustafsson, *Control of error and convergence in ODE solvers,* Ph.D. thesis, Department of Automatic Control, Lund Institute of Technology, Sweden, 1992.

[91] K. Gustafsson, M. Lundh, and G. Söderlind, *A PI stepsize control for the numerical solution of ordinary differential equations,* BIT **28** (1988), 270–287.

[92] W. Hackbusch, *Elliptic Differential Equations,* Springer, 1992.

[93] E. Hairer, S. P. Nørsett, and G. Wanner, *Solving ordinary differential equations. I: Nonstiff problems,* 2nd rev. ed., Springer Series in Computational Mathematics, vol. 8, Springer, Berlin, 1993.

[94] E. Hairer and G. Wanner, *Solving ordinary differential equations. II: Stiff and differential–algebraic problems,* 2nd rev. ed., Springer Series in Computational Mathematics, vol. 14, Springer, Berlin, 1996.

[95] H. Harbrecht and R. Stevenson, *Wavelets with patchwise cancellation properties,* Math. Comput. **75** (2006), 1871–1889.

[96] W.H. Hundsdorfer, J.M. Sanz-Serna, and J.G. Verwer, *Convergence and order reduction of Runge-Kutta schemes applied to evolutionary problems in partial differential equations,* Numer. Math. **50** (1987), 405–418.

[97] S. Jaffard, *Propriétés des matrices "bien localisées" près de leur diagonale et quelques applications,* Ann. Inst. H. Poincaré Anal. Non Linéaire **7** (1990), no. 5, 461–476.

[98] D. Jerison and C.E. Kenig, *The inhomogeneous Dirichlet problem in Lipschitz domains,* J. Funct. Anal. **130** (1995), no. 1, 161–219.

[99] R. Q. Jia and Q. T. Jiang, *Spectral analysis of the transition operator and its applications to smoothness analysis of wavelets,* SIAM J. Matrix Anal. Appl. **24** (2003), no. 4, 1071–1109.

[100] R. Q. Jia and S. T. Liu, *Wavelet bases of Hermite cubic splines on the interval,* Adv. Comput. Math. **25** (2006), no. 1–3, 23–29.

[101] H. Johnen and K. Scherer, *On the equivalence of the K-functional and moduli of continuity and some applications*, Constr. Theory Funct. several Variables, Proc. Conf. Oberwolfach 1976, Lect.Notes Math., vol. 571, Springer, 1977, pp. 119–140.

[102] M. Jürgens, *Adaptive Wavelet-Verfahren auf allgemeinen Gebieten*, Diplomarbeit, RWTH Aachen, 2001.

[103] _____, *A semigroup approach to the numerical solution of parabolic differential equations*, Ph.D. thesis, RWTH Aachen, 2005.

[104] J. Kadlec, *On the regularity of the solution of the Poisson problem on a domain with boundary locally similar to the boundary of a convex open set*, Czech. Math. J. **14** (1964), no. 89, 386–393.

[105] T. Kato, *Perturbation theory for linear operator equations*, second ed., Grundlehren der mathematischen Wissenschaften, vol. 132, Springer, 1976.

[106] A. Kunoth and J. Sahner, *Wavelets on manifolds: An optimized construction*, Math. Comput. **75** (2006), no. 255, 1319–1349.

[107] J. Lang, *Adaptive multilevel solution of nonlinear parabolic PDE systems. Theory, algorithm, and applications*, Preprint SC 99–20, Konrad–Zuse–Zentrum für Informationstechnik Berlin, 1999.

[108] _____, *Adaptive multilevel solution of nonlinear parabolic PDE systems. Theory, algorithm, and applications*, Lecture Notes in Computational Science and Engineering, vol. 16, Springer, Berlin, 2001.

[109] J. Lang and J.G. Verwer, *ROS3P — An accurate third-order Rosenbrock solver designed for parabolic problems*, BIT **41** (2001), 731–738.

[110] _____, *On global error estimation and control for initial value problems*, Preprint MAS–E0531, Centrum voor Wiskunde en Informatica, Amsterdam, 2005.

[111] S. Lang, *Introduction to complex hyperbolic spaces*, Springer, New York, 1987.

[112] I. Lasiecka, *Unified theory for abstract parabolic boundary problems — a semigroup approach*, Appl. Math. Optimization **6** (1980), 287–333.

[113] P. G. Lemarié, *Bases d'ondelettes sur les groupes de Lie stratifiés*, Bulletin de la Société Mathématique de France **117** (1989), no. 2, 213–232.

[114] C. Lubich and A. Ostermann, *Linearly implicit time discretization of nonlinear parabolic equations*, IMA J. Numer. Anal. **15** (1995), no. 4, 555–583.

[115] A. Lunardi, *Analytic semigroups and optimal regularity in parabolic problems*, Progress in Nonlinear Differential Equations and Their Applications, vol. 16, Birkhäuser, 1995.

[116] S. Mallat, *Multiresolution approximation and wavelet orthonormal bases of* $L_2(\mathbb{R}^d)$, Trans. Amer. Math. Soc. **315** (1989), 69–87.

[117] Y. Meyer, *Ondelettes, fonctions splines et analyses graduées*, Rend. Semin. Mat. Univ. Politec. Torino **45** (1987), 1–42.

[118] ———, *Ondelettes et opérateurs I: Ondelettes*, Hermann, 1990.

[119] P. Morin, R. H. Nochetto, and K. G. Siebert, *Convergence of adaptive finite element methods*, SIAM Rev. **44** (2002), no. 4, 631–658.

[120] A. Ostermann and M. Roche, *Runge–Kutta methods for partial differential equations and fractional orders of convergence*, Math. Comput. **59** (1992), no. 200, 403–420.

[121] ———, *Rosenbrock methods for partial differential equations and fractional orders of convergence*, SIAM J. Numer. Anal. **30** (1993), no. 4, 1084–1098.

[122] P. Oswald, *Multilevel finite element approximation. Theory and applications.*, Teubner Skripten zur Numerik, Teubner, Stuttgart, 1994.

[123] ———, *Frames and space splittings in Hilbert spaces*, Survey lectures on multilevel schemes for elliptic problems in Sobolev spaces. http://www.faculty.iu-bremen.de/poswald/bonn1.pdf, 1997.

[124] ———, *Multilevel frames and Riesz bases in Sobolev spaces*, Survey lectures on multilevel schemes for elliptic problems in Sobolev spaces http://www.faculty.iu-bremen.de/poswald/bonn2.pdf, 1997.

[125] A. Pazy, *Semigroups of linear operators and applications to partial differential equations*, Applied Mathematical Sciences, vol. 44, Springer, 1983.

[126] M. Primbs, *Stabile biorthogonale Spline–Waveletbasen auf dem Intervall*, Dissertation, Universität Duisburg–Essen, 2006.

[127] M. Roche, *Rosenbrock methods for differential algebraic equations*, Numer. Math. **52** (1988), 45–63.

[128] J.M. Sanz-Serna and J.G. Verwer, *Stability and convergence at the PDE/stiff ODE interface*, Appl. Numer. Math. **5** (1989), no. 1/2, 117–132.

[129] R. Schneider, *Multiskalen– und Wavelet–Matrixkompression. Analysisbasierte Methoden zur effizienten Lösung großer vollbesetzter Gleichungssysteme*, Habilitationsschrift, TH Darmstadt, 1995.

[130] D. Sheen, I.H. Sloan, and V. Thomée, *A parallel method for time–discretization of parabolic problems based on contour integral representation and quadrature*, Math. Comput. **69** (1999), no. 229, 177–195.

[131] ———, *A parallel method for time–discretization of parabolic equations based on laplace transformation and quadrature*, IMA J. Numer. Anal. **23** (2003), no. 2, 269–299.

[132] G. Steinebach, *Order-reduction of ROW-methods for DAEs and method of lines applications*, Tech. Report 1741, TH Darmstadt, 1995.

[133] R. Stevenson, *Adaptive solution of operator equations using wavelet frames*, SIAM J. Numer. Anal. **41** (2003), no. 3, 1074–1100.

[134] _____, *On the compressibility of operators in wavelet coordinates*, SIAM J. Math. Anal. **35** (2004), no. 5, 1110–1132.

[135] _____, *Optimality of a standard adaptive finite element method*, Found. Comput. Math. (2006), published online, DOI 10.1007/s10208-005-0183-0.

[136] _____, *Composite wavelet bases with extended stability and cancellation properties*, SIAM J. Numer. Anal. **45** (2007), no. 1, 133–162.

[137] K. Strehmel and R. Weiner, *Linear–implizite Runge–Kutta–Methoden und ihre Anwendung*, Teubner, Leipzig, 1992.

[138] H. Tanabe, *Equations of evolution*, Pitman, London, 1979.

[139] H. Triebel, *Interpolation theory, function spaces, differential operators*, North–Holland, Amsterdam, 1978.

[140] _____, *Theory of function spaces*, Birkhäuser, Basel/Boston/Stuttgart, 1983.

[141] M. Werner, *Adaptive wavelet frame domain decomposition methods for elliptic problems*, Dissertation, Philipps–Universität Marburg, in preparation.

[142] _____, *Adaptive Frame-Verfahren für elliptische Randwertprobleme*, Diplomarbeit, Fachbereich Mathematik und Informatik, Philipps-Universität Marburg, 2005.

[143] P. Wojtaszczyk, *A mathematical introduction to wavelets*, Cambridge University Press, 1997.

Lebenslauf

Persönliche Daten

Name:	Thorsten Joachim Raasch
Geburtsdatum:	3. Dezember 1975
Geburtsort:	Frankenberg/Eder
Staatsangehörigkeit:	deutsch
Familienstand:	ledig

Schulausbildung

07/1982 – 06/1986	Grundschule Züschen
07/1986 – 06/1995	Geschwister-Scholl-Gymnasium der Stadt Winterberg
06/1995	Allgemeine Hochschulreife

Wehrdienst

10/1995 – 09/1996

Studium

10/1996 – 03/2001	Studium der Mathematik mit Nebenfach Wirtschaftswissenschaften, Universität Siegen
09/1998	Vordiplom im Studienfach Mathematik
03/2000 – 03/2001	zusätzlich Studium der Technischen Informatik, Universität Siegen
03/2001	Diplom im Studienfach Mathematik

Berufliche Tätigkeit

04/1999 – 03/2001	studentische Hilfskraft, AG Angewandte und Numerische Mathematik, Prof. Dr. H.-J. Reinhardt, FB Mathematik, Universität Siegen
03/2001 – 04/2001	wissenschaftliche Hilfskraft mit Abschluss, AG Angewandte und Numerische Mathematik, Prof. Dr. H.-J. Reinhardt, FB Mathematik, Universität Siegen
04/2001 – 04/2002	wissenschaftlicher Mitarbeiter, AG Angewandte und Numerische Mathematik, Prof. Dr. H.-J. Reinhardt, FB Mathematik, Universität Siegen
04/2002 – 03/2007	wissenschaftlicher Mitarbeiter, AG Numerik/Wavelet–Analysis, Prof. Dr. S. Dahlke, FB Mathematik und Informatik, Philipps–Universität Marburg